GLUTEN-FREE FLAVOR FLOURS

Also by ALICE MEDRICH

Cocolat: Extraordinary Chocolate Desserts
Chocolate and the Art of Low-Fat Desserts
Alice Medrich's Cookies and Brownies
A Year in Chocolate: Four Seasons of Unforgettable Desserts
Chocolate Holidays
Pure Dessert
Chewy Gooey Crispy Crunchy Melt-in-Your-Mouth Cookies
Sinfully Easy Delicious Desserts
Seriously Bitter Sweet

ALICE MEDRICH

WITH MAYA KLEIN

GLUTEN-FREE FLAVOR FLOURS

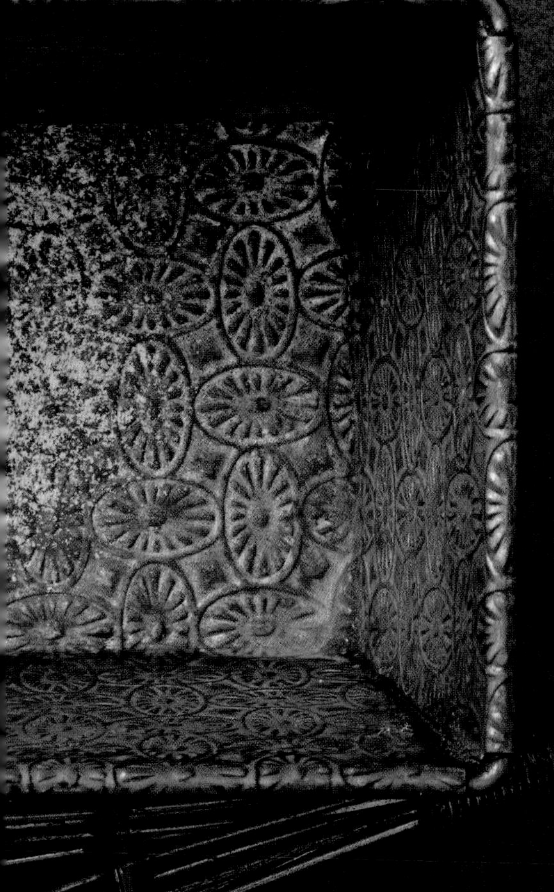

A New Way to Bake with Non-Wheat Flours

INCLUDING
Rice, Nut, Coconut, Teff, Buckwheat, and Sorghum Flours

ARTISAN • NEW YORK

Library of Congress Cataloging-in-Publication Data
Names: Medrich, Alice, author. | Klein, Maya, author.
Title: Gluten-free flavor flours / Alice Medrich with Maya Klein.
Other titles: Flavor flours
Description: New York, NY : Artisan, [2017] | Revision of: Flavor flours. | Includes bibliographical references and index.
Identifiers: LCCN 2017002004 | ISBN 9781579658069 (pbk. : alk. paper)
Subjects: LCSH: Flour. | Gluten-free diet—Recipes. | Baking. | LCGFT: Cookbooks.
Classification: LCC TX393 .M43 2017 | DDC 641.3/31—dc23 LC record available at https://lccn.loc.gov/2017002004

Originally published in hardcover as *Flavor Flours* by Artisan in 2014.

Front cover photograph by Lauren Volo
Back cover photographs by Leigh Beisch

Design by Laura Klynstra

Photo styling and creative direction by Sara Slavin
Food styling by Sandra Cook

Published by Artisan
A division of Workman Publishing Co., Inc.
225 Varick Street
New York, NY 10014-4381
artisanbooks.com

Published simultaneously in Canada by Thomas Allen & Son, Limited

Printed in China

10 9 8 7 6 5 4 3 2 1

To our children
Lucy Medrich
Nate Klein
Wade Klein

CONTENTS

INTRODUCTION 11

CHAPTER ONE:
RICE FLOUR
51

CHAPTER TWO:
OAT FLOUR
83

CHAPTER FIVE:
CHESTNUT FLOUR
197

CHAPTER SIX:
TEFF FLOUR
229

CHAPTER THREE:
CORN FLOUR AND CORNMEAL
137

CHAPTER FOUR:
BUCKWHEAT FLOUR
167

CHAPTER SEVEN:
SORGHUM FLOUR
261

CHAPTER EIGHT:
NUT AND COCONUT FLOURS
295

CHAPTER NINE: ELEMENTS 333

RESOURCES 351

APPENDIX 353

ACKNOWLEDGMENTS 357

INDEX 358

INTRODUCTION

Whether or not you are gluten-intolerant, if you are an avid cook or baker, or someone who tinkers in the kitchen for fun, you probably walk the aisles of supermarkets and haunt specialty food shops for new and interesting ingredients to taste and experiment with. In addition to all-purpose gluten-free flour blends, you've no doubt noticed the rainbow of so-called alternative flours a few steps from the "regular" flour in the baking aisle. Oat flour, brown rice flour, corn flour, even non-grain flours like buckwheat and all kinds of nut flours . . . Maybe you've tried some of them. Or maybe you still walk by, but wonder what to do with them. I stopped passing them by and started experimenting with a few of the most flavorful of these gorgeous flours several years ago for my book *Pure Dessert*. My strategy then was to replace a portion of the wheat flour in a recipe with a new flour to create a more interesting and flavorful cake or cookie. In essence, I treated the new flours like flavor ingredients that also happened to have some of the bulky and absorbent characteristics of the flour (wheat) that we are most familiar with. The results were variously delicate, interesting, elegant, and decadent—all things desserts should be—but they also made me smile to know that whole grains could be added to desserts because they tasted good rather than because they were healthy! "Two birds with one stone," I thought, and also, "I'm not done here."

Pure Dessert was just the beginning. Flavorful flours continued to beckon, and there are even more of them now (thanks to the growing number of gluten-free eaters). Now everyone can find these flours in better supermarkets across the country, as well as online. Meanwhile, people who love food are more interested in and open to the idea of flours that taste and feel different from the all-purpose wheat flour standard. After a terrific response to the wheat-free cookies in my book *Chewy Gooey Crispy Crunchy Melt-in-Your-Mouth Cookies*, I was excited about working more with "flavor flours," treating them as "hero" ingredients (not just substitutions), and challenged by the idea of using them exclusively—without any wheat flour in an adventurous new kind of gluten-free baking.

Gluten-Free Flavor Flours is a collection of recipes that feature individual gluten-free flours: rice, oat, corn, sorghum, teff, buckwheat, coconut, and chestnut and other nuts. I chose these flours because each has a very distinct flavor—a voice, really. With the arguable exception of white rice flour, flours that are essentially pure starch are not included, partly because they aren't that delicious and don't have much flavor, and partly because I love bringing more whole grains into baking, especially into decadent and indulgent desserts, where they are still unexpected.

Rice, oats, and corn (and to a limited extent, buckwheat) are familiar to us as a side dish, a bowl of porridge or groats, or a hearty bread. But once these familiar grains are transformed into flours, they can be used in baked goods that have entirely different textures. In the 1970s chefs started serving vegetable purees, many of which were hard to identify. Why? Aside from the fact that tons of butter or cream was added to them, we no longer had a familiar texture for reference. We had to pay attention. Suddenly, we tasted flavors that we'd never noticed before. Tasting familiar ingredients in new forms likewise reveals unexpected flavors. Grains like brown rice and oats have a known pedigree that connects them to health food, hearty bread, and granola, so I was startled at the delicate textures and aromas I found in the simplest cakes made from their flours. I never dreamed that a plain oat flour sponge cake would taste like butterscotch, or that a brown rice sponge could have such a moist and delicate butter flavor, or that buckwheat would have notes of honey and rose.

Teff and sorghum are consumed around the world, but rarely by Americans, so I had no experience baking with them but plenty of excitement about exploring the possibilities. Rather than use them as they are used indigenously, I treat them like new ingredients in familiar rather than ethnic baking. Teff is earthy and sometimes tastes very much like whole wheat but without the bitterness of whole wheat. Sorghum has flavors reminiscent of oats and corn, yet is somehow different. Chestnuts are most familiar as a wintertime treat, roasted and sold on street corners, or used in traditional turkey stuffing. The flour produced from dried chestnuts is soft and sweet, a wonderful and inspiring baking ingredient. Rice flour has its own distinct flavor and can be used by itself, but it also works as a neutral backdrop for other ingredients.

Nuts and coconut are common in Western baking; every good baker's notebook includes a variety of European nut torten made with ground nuts or nut meal—now often called nut flour. Used as the major ingredient in recipes without any other flour or as a partner to other flavor flours, nut flours make sensational cakes, cookies, and more. Coconut, once a polarizing ingredient (people either loved it or hated it), is now very much in vogue. The easy availability of coconut flour and dried coconut—not just the old supersweet shreds—makes coconut a more versatile and compelling ingredient than ever. I have always loved coconut and now have a reason to include it as a major ingredient rather than just a flavor accessory, and an excuse for so many coconut cookies in one book!

Gluten-Free Flavor Flours is organized by flour; each chapter includes recipes that call for that flour alone, as well as recipes with a partner flour that supports it. In conventional gluten-free recipes, specialty flours like those used here are mixed with other flours and starches to create an all-purpose neutral blend that can be used as a substitute for wheat flour in everyday baking. I've taken a different, more ingredient-driven approach here, by letting each flour star in a variety of delicious and surprising recipes—as though wheat never existed! Blends of several flours are rare and are used expressly to build bold and complex flavor profiles (as is the case in three sensational savory crackers), rather than to create a neutral background. Since the goal of this book is to celebrate the flavor of the flours, each one is treated as a hero, or important feature of the recipe; I do not use an all-purpose flour blend.

Most of the recipes are for beloved familiar desserts and other baked goods: brownies and chocolate cakes, biscuits and quick breads, cookies, doughnuts, and crackers. In each recipe, the flour is used for its own flavor and sometimes also to enhance the ensemble of flavors in the recipe. Flours with nutty flavors like teff and buckwheat add complexity and complementary flavor to a nutty fruitcake. Brownies made with rice flour taste more chocolaty than "regular" brownies because the rice flour seems to amplify the flavor of the chocolate! The same is true of chocolate soufflés and many other desserts. Brownies made with teff flour get an added nuance of cocoa flavor from the teff. The deep, earthy flavor of buckwheat is such a natural in Panforte Nero (page 184), with its warm spices, black figs, and amber honey, it's probably hard to imagine that Buckwheat Sour Cream Soufflés (page 193) drizzled with honey could be so delicate and floral by contrast. A classic Swiss tart filled with honey and walnuts gets bonus flavor from a chestnut flour crust, as does a ricotta cheesecake; chestnut génoise is good enough to nibble with your coffee all by itself, and even better with crème fraîche and spiced pear or apple butter.

Rice flour alone is spectacularly delicious in a chiffon cake—you will be amazed—but in biscuits and some butter cakes, the flavor is more nuanced and complex with a little oat flour blended in.

The "new" flours do require some new rules and techniques, but surprisingly few. Most of the more fiddly, intricate techniques of classical baking—designed to outfox gluten and prevent it from producing tough cakes and cookies—are no longer necessary. Mixing is done in fewer steps with less fuss. Baking in a tube pan or a sheet pan (or two or three shallow pans) rather than a regular deep cake pan gives support to some of the most delicate cakes; a little xanthan gum and plenty of mixing provide needed structure—and a velvety texture—for other types of cakes. In a sense, the most surprising aspect of the "new" techniques may be the extent to which some of the old techniques are no longer necessary.

Exploring with these fantastically flavorful ingredients has been the continuation of a lifelong adventure and joyful collaboration in the kitchen. You will discover, as I have, that these flours and flavors belong in every passionate baker's repertoire. I hope these new recipes and fresh ideas add excitement and variety to your table and elicit smiles of pleasure from your guests. I hope you build on them and create new traditions as well. I wish you many more hours of happy and rewarding baking.

THE PROCESS

I knew that a book like this one—about new ingredients and new ways of using them—would require endless experimentation, leaps of imagination, creativity, and sometimes the ability to hold two opposing ideas long enough to discover some intricate reasoning that might make both of them true. There were so many new questions—with every answer seemingly raising at least five more questions—that I knew I needed a partner.

I enlisted Maya Klein to join the project. Portland, Oregon–based cooking teacher, culinary consultant, and great friend, Maya has worked with me for decades—starting with recipe testing when I revised several of the baking chapters for the 1997 *New Joy of Cooking*. Maya continued to work with me on my own books, testing and creating recipes, contributing ideas, and talking through problems. I don't know anyone else with her palate and her ability to combine intuition and creativity with analysis—or with whom hard questions are so much fun to ponder. Meanwhile, Maya is herself wheat sensitive and has been reinventing her own baking for several years. She immediately "got" the idea of baking with flavorful flours in a new way rather than treating them as wheat flour substitutes, and her immediate reaction turned into a mantra: What would baking be like if wheat didn't exist?

Our ongoing dialogue was invaluable, as always, but this time it proved essential: the process of baking with mostly unfamiliar flours was filled with unexpected challenges and surprising outcomes. Sometimes, for example, we were both working with the same flour but in different types of recipes and getting wildly different flavors and textures—it seemed as though we were each seeing a different part of the same elephant. But this forced us to figure out why we were getting different results and how we could use the lesson to our advantage. The story about buckwheat testing and the lessons we learned about using it (see page 168) just begin to capture the benefits of a great collaboration. What we learned and produced together was far greater than the sum of our individual talents and experience. If you want to understand our testing process or are wondering how the recipes in *Gluten-Free Flavor Flours* differ from traditional wheat-based baking, read on; our thinking and experimenting are described and summarized in the next few sections as well as in the recipes.

What If Wheat Didn't Exist?

Western bakers are so accustomed to wheat flour that its behavioral quirks are simply taken for granted. We hardly notice that traditional baking techniques are built around wheat, specifically designed to make gluten work for us by either inhibiting or encouraging its development, depending on whether the goal is a tender cake or a chewy bread. But what if wheat flour did not exist?

None of the flavor flours in this book behaves like wheat flour, nor do they necessarily behave like one another. Some but not all are grains, and even the grains differ in texture. One of the flours comes from a nut but behaves more like a starch—and the other nut flours are a completely different story. How could these distinctive flours be turned into delicious cakes, cookies, biscuits, and other baked goods? What tricks and techniques (brand-new, classic, or from the gluten-free playbook) would let them shine, rather than just "substitute" for wheat? How could some of the most egregious problems (a gritty, pasty, gummy, or crumbly texture or a raw-flour flavor) of wheat-free baking be avoided? Would it be necessary to add xanthan gum or seed meal to every recipe to bind the ingredients and provide structure? Was it even possible to use single flours in recipes rather than blends?

Our knowledge of classical baking certainly gave us an arsenal of techniques and tricks to make the new flours work, and we borrowed selectively from gluten-free baking, but new thinking was required, too. Considering the hundreds of hours we spent testing and experimenting, most of the final recipes are simple to execute, even for a beginning baker.

DISCOVERIES AND LESSONS FROM A GÉNOISE

At the outset of this project, I wondered if the flours that I had chosen for their flavors could be made into a cake of any kind without adding a gluten substitute such as xanthan gum or seed meal. It seemed that the easiest way to jump in and get acquainted with the flour flavors, textures, and general "behavior" would be to bake the same recipe with each of the flours individually, and just see what happened. I chose a simple recipe for génoise, so easy I could practically do it in my sleep. Génoise is *the* classic sponge or foam cake and has only four ingredients plus salt. I chose it because it gets its leavening and structure from whipped whole eggs, the proportion of which (I thought/hoped) might provide enough structure to make a reasonable cake.

Now, the classic génoise is a light but fairly dry cake, one often used in multilayered desserts where the layers are first moistened and flavored with some syrup before being filled and frosted. Although I have always prided myself on making a very tasty one, génoise is not normally considered moist enough or flavorful enough to stand alone.

As it turned out, every one of those first génoise samples made with flavor flours was at least promising: some were incredibly delicious just nibbled with a cup of coffee; others needed a little more or less butter or flour, the addition of egg yolks, a switch to oil, or a different pan or mixing detail. Each tasted of the flour that was used to make it in the nicest possible way and each sparked ideas for new flavor combinations. There was a bonus: I ended up with a stack of sponge cakes in an array of gorgeous natural colors—creamy white, buff, tan, goldenrod, and gray brown. I was already imagining the photos for the book.

The génoise exercise confirmed that it was possible to work with the flours individually, without the need to always blend them or to add a gum such as xanthan or guar for structure. It was an important reminder of how eggs provide structure to a cake, especially when they are whipped into a foam. Génoise was an encouraging start, but it was not an accurate predictor of success, because other types of recipes turned out to be trickier. The génoise was just the beginning of this journey.

You will find a génoise (or sponge cake) in every chapter. Some are moist and interesting enough to be eaten plain or with a little topping or fruit accompaniment; others need a moist filling or a splash of flavorful syrup plus frosting. The most neutral of all (and arguably the least interesting) are made with white rice or sorghum; but either could substitute perfectly for a classic plain génoise in the French baking repertoire. Pastry chefs can start with these recipes, whether or not they are baking for a gluten-free audience. You will also find a chocolate génoise made with teff flour (see page 232) that is an improvement on the classic. Meanwhile, sponge cakes made with rice, oat, corn, buckwheat, chestnut, sorghum, and nut flours—found at the beginning of each chapter—will dramatically expand the repertoire of any passionate home baker or pastry chef.

Chiffon cake testing turned out to be another happy surprise. Look for a gluten-free chiffon cake recipe online and you will find ones that call for a blend of three or four flours and starches plus xanthan gum—the whole nine yards of gluten-free baking. I was excited to discover that an astonishingly delicious chiffon cake can be made with pure white rice flour, a combination of brown and white rice flours, or a combination of corn flour and white rice flour, without any xanthan gum at all. There is one big problem with chiffon cakes, though: they are too delicate to cool upside down in the traditional manner. Several cakes had to fall out of their pans before the lightbulb went on: stop turning those cakes upside down in the first place!

While I was finding that sponge cakes could be made without xanthan gum, Maya was reinventing carrot cake. Génoise and carrot cake would not seem to have anything in common, but if the eggs, oil, and sugar in the latter are whipped to a thick, stable foam (a bit like for the génoise), rather than just mixed together as they normally are, the carrot cake gets additional structure. Carrot cake batter also needs baking powder because it is heavier than génoise batter, but it doesn't need xanthan gum.

Gingerbread, biscuits, doughnuts, and crackers were even trickier to master. It was necessary to get the right balance of wet and dry ingredients and to make sure that the flour was hydrated (had absorbed enough moisture). Hydration is often cited as a critical factor in successful non-wheat-based baking, and it is more complicated than it sounds because each flour is different, and each type of recipe works differently. There were plenty of ifs, ands, and buts and no one solution to fit them all.

Tart and cookie doughs and cookie batters presented another obstacle. To improve flavor and prevent gritty or powdery textures, we found that these usually needed a little more liquid than usual, as well as resting time to let the doughs fully hydrate. Cream cheese helped to bind butter cookies and tart crust ingredients and provided some structure. True confession: although we did not add xanthan gum to these recipes, I suspect that the gum in commercial cream cheese provides some of the same function!

Crackers added an extra complication. These doughs did need extra liquid and hydration to prevent grittiness and the flavor of uncooked grains—but overnight resting made the nuts and seeds rubbery, resulting in crackers that were not fully crisp and crunchy. The solution was a wetter-than-usual dough (with no rest), rolled thin and baked at a temperature high enough to simultaneously cook the flour fully and drive out all of the moisture. Voilà, super-crunchy crackers.

Butter cake batters with plenty of liquid and lots of mixing raised another challenge. It seems intuitive that too much liquid might produce a gummy cake and too little a dry, crumbly cake. But even the "perfect" amount of liquid makes rubbery or gummy cakes if the ingredients are not mixed in the right order. If you bake, you might recall that some classic (wheat flour) butter cake methods call for mixing the flour with the butter before the wet ingredients are added. This is done to waterproof the flour particles and prevent excess gluten development (which would result in a tough cake) when the liquid is beaten in. Sans gluten, we assumed that waterproofing the flour was not necessary. But protecting the flour with fat *did* turn out to be necessary sometimes, not for outsmarting gluten (there is none), but for preventing the starches in the flour from *over*hydrating too early and thereby causing a gummy texture. Go figure.

Solutions to structural problems were sometimes hidden in plain sight: some cakes are too delicate, just not strong enough, to bake in a deep pan without sinking a little in the center. The answer was hard-won but so simple: change the pan rather than the recipe. Tube pans are used to support some cakes; others are baked in a wider, shallower pan to make a thin cake sheet, which can be rolled up to make a roulade or cut into layers and stacked with filling.

While it was useful to borrow the simple techniques for coating flour with fat or whipping eggs and sugar (and fat) to a stable foam, the most finicky techniques of classic baking can be entirely ignored with these flours: ingredients usually do not have to be at room temperature, flour need not ever be added to batters in three parts alternating with two parts of liquid, and overmixing is rarely the problem because absent gluten, mixing cannot make a cake or biscuit tough.

The ideal biscuit is shapely, golden, and slightly craggy, with a soft and tender crumb, a lovely crusty exterior, and plenty of buttery flavor—and you'll find all of that in the recipe on page 109. But the early results ran the gamut of flaws: flat and firm (too much flour); flat and soft (too much liquid and not enough xanthan gum); coarse, uneven crumb (butter instead of cream); soft crumb but collapsed out of the oven (underbaking at the wrong temperature and/or undermixing); or one-dimensional flavor (just cream, no yogurt for tanginess).

The doughnut experience flies in the face of the common view that frying makes anything taste good. It doesn't. Fried things can be soggy, greasy, and uninteresting. But once the recipe works, rice flour doughnuts have an extraordinary and delicious advantage: the flavors of the butter and eggs come forward because of the delicacy of the rice flour. Moreover, the glutinous rice flour keeps the doughnuts soft and chewy so they can be made in advance and reheated to their original glory, then dusted with sugar or drizzled with chocolate as though freshly fried. This is a huge advance for any cook who hates to fry in the presence of company or kids, or when wearing party duds.

As you might expect, recipes with very little flour are a natural for flavor flours. You will find great variety among them, from dense fruit- and nut-laden cakes and confections, such as Date-Nut Cake with Apricots and Teff (page 244) and Panforte Nero (page 184), to the lightest, airiest soufflés and rich chocolate tortes. The flour in these recipes is only a minor source of texture (and structure), but it's an important flavor partner for the other ingredients, or, in the case of some of the soufflés, *the* featured flavor.

WHAT TO EXPECT IN THE RECIPES

To make the recipes in this book successfully, you do not need to know any of the details of the testing or problem solving (described above) that went into creating them, or even that the recipes that were the trickiest to develop are probably the easiest to make. But here is an overview of techniques and what to expect as you work through this book.

The various flours work differently in different types of recipes: you will find several mixing techniques and many riffs and variations.

Most recipe techniques will seem familiar from regular (wheat-based) baking, but they may be less involved or finicky: traditional methods, sometimes streamlined or adapted, were used where they worked and discarded where they didn't.

Most of the recipes—including butter cakes, biscuits and scones, and muffins—are mixed in one or two simple steps with an electric mixer. They are easy to make and practically foolproof. Don't skimp on mixing time when times are given if you want to get the best possible results.

Don't expect butter cake layers and cupcakes to dome on top: flours without gluten are not strong enough to cause doming. This is an advantage for layer cakes: it's easy to make professional-looking frosted cakes because there is no need to trim the domes to make them level or fill them with flat sides together when you stack and frost them. If you miss shapely domed cupcakes, you'll just have to get over it, or compensate with frosting. If you are a frosting minimalist, heap it atop the cupcake in the center rather than at the edges to create a graceful (or perky) contour. If you *do* love frosting, just double the amount and mound or pipe it all over, sky-high!

Sponge cakes and those that involve whipped egg whites are slightly more challenging than other cakes; most require hand folding because batters are light and delicate. But the flours are easier to fold in than wheat flour, so sponge batters are less apt to deflate when mixed. Sponge cakes may sink slightly as they cool; recipes may call for you to trim the edges to level the cake before cutting it into layers for filling.

Most of the flours need some amount of hydration to prevent the finished product from feeling or tasting gummy, powdery, gritty, or even (from experience) like sandpaper, but too much hydration too early can be a problem. The recipes forestall this concern: batters are mixed in such a way, or rested, to ensure that the flour has time to absorb the right amount of moisture at the right time. You don't have to think about this.

Xanthan gum, which is widely used in gluten-free baking to bind ingredients and to create structure (so that cakes have a nice crumb or cookies have a little chewiness), is only necessary in some recipes and usually in smaller amounts than called for in gluten-free recipes elsewhere.

Ground flaxseed or other seed meals (often used as a binder in non-wheat-based batters) are never used as a binder here, unless their very distinctive flavors are important to the flavor profile of the finished product. Seed Crackers (page 163), Walnut and

Buckwheat Crackers (page 194), and Tangy Aromatic Crackers (page 259) are stunning examples.

CREATING YOUR OWN RECIPES WITH FLAVOR FLOURS

It would be fantastic if there were rules for taking any recipe that calls for ordinary all-purpose wheat flour and converting it for any other flour. But it's not that straightforward.

The various flours work differently in different types of recipes and are different from one another. Some flours are soft and prone to mushiness in one type of recipe and perfect in another. Flours that are gritty in one context are divinely crisp and crunchy in another. Certain techniques bring out the best in these flours, but the techniques are not the same for every flour. It took hundreds of hours of experimentation to make the recipes in this book. My best advice is to make the recipes first as written, and if you like the results, then experiment further. If you want a light cake, riff off one of the génoise or sponge cakes. If you want a cake with the texture of a butter cake or pound cake, but with different flavors or ingredients than those given here, use the Ultimate Butter Cake (page 90) as a starting point. For a new cracker, don't start with a wheat flour cracker recipe and try to convert it; start with one of these cracker recipes (see pages 163, 194, and 259) and switch out the elements until you have something that approaches your vision.

FLAVORFUL PANCAKES, WAFFLES, AND CREPES

The preceding caveats about trying new flours don't hold true for most pancakes, waffles, and crepes. Because they are thin—baked right on a griddle or waffle iron rather than in deep cake pans—they don't require any more structure than the usual number of eggs can provide, and they already contain plenty of liquid to hydrate the flour. All of this means that pancakes, waffles, and crepes are perfect for experimentation and for introducing flavor flours into your normal repertoire. You can substitute most of the flours—white and brown rice, oat, corn, buckwheat, chestnut, teff, and sorghum—for all-purpose flour by either volume or weight and adjust the consistency of the batter by adding more liquid or flour, as one does anyway with this type of recipe. You can do the same with dried shredded coconut or any nut flour (other than chestnut) if you also add half again as much rice flour: results will be pleasingly dense, slightly chewy, super flavorful, and completely irresistible. (Coconut flour is more complicated and not addressed here.) You can also blend flours to get different flavor effects, and of course you can add all of the chopped nuts, seeds, chocolate chips, or berries that you might already mix into your pancakes and waffles. I've included basic recipes here, but if you have favorite family recipes, chances are they, too, will work perfectly and deliciously with your new flours.

Beyond Maple Syrup

Maple syrup is always good—but here are some additional flavorful toppings and accompaniments to complement pancakes and waffles made with flavorful flours.

Brown Rice Flour Pancakes or Waffles

- Fresh strawberries or blueberries
- Dried cranberries or blueberries added to the batter
- Dulce de leche and crumbled cotija cheese

Oat Flour Pancakes or Waffles

- Maple syrup with bananas and walnuts or toasted pecans
- Blueberry syrup or fresh blueberries and powdered sugar
- Strawberries or bananas with sour cream and a sprinkling of dark brown or (better yet) dark muscovado sugar
- For a crunchy, flavorful crust on waffles, sprinkle chopped walnuts or pecans over the batter before closing the waffle iron.

Corn Flour Pancakes or Waffles

- Berry preserves or berry syrups
- For pancakes: Sprinkle a few fresh or frozen blueberries onto the batter on the griddle.

Buckwheat Flour Pancakes or Waffles

- Sour cream or crème fraîche and cherry or berry preserves
- Sour cream or crème fraîche and a drizzle of honey or sorghum syrup

Chestnut Flour Pancakes or Waffles

- Ricotta cheese and fresh figs with a drizzle of honey
- Spicy applesauce or apple or pear butter and sour cream or crème fraîche
- Sour cream and honey or cherry preserves

Sorghum Flour Pancakes or Waffles

- Blackberry preserves
- Sorghum syrup

Nut Flour Pancakes or Waffles

- Chocolate chips in the batter
- Berry preserves with ricotta cheese or mascarpone
- Peaches or berries and whipped cream
- Chopped dates and Greek yogurt with vanilla syrup (such as Torani or Monin brand)

Coconut Pancakes or Waffles

- Mango or papaya slices with a squeeze of lime, palm sugar, or brown sugar
- Plain or grilled pineapple fans sprinkled with dark muscovado sugar or brown sugar
- Drizzle of dulce de leche

BASIC PANCAKES

MAKES ABOUT FIFTEEN 4-INCH PANCAKES

1¼ cups flour: white or brown rice, gluten-free oat, corn, buckwheat, chestnut, teff, sorghum, or dried shredded coconut, or nut flour/meal (see pages 353–354 for equivalent weights for each flour)

If using shredded coconut or a nut flour other than chestnut, add ½ cup plus 2 tablespoons (95 grams) white rice flour

1 tablespoon sugar

2 teaspoons baking powder

Slightly rounded ½ teaspoon salt

2 large eggs

4 tablespoons (½ stick/55 grams) unsalted butter, melted and warm, plus more for the skillet

1 cup milk, warm

Whisk the flour(s), or coconut or nut flour and rice flour, with the sugar, baking powder, and salt in a medium bowl. Add the eggs, butter, and milk and whisk until smooth and well blended. Add a little more milk or flour at any time if the batter seems too thick or too thin.

Heat a large skillet or griddle over medium heat until hot and brush it with butter. Pour 4-inch pancakes (a generous tablespoon) and cook for 1 to 2 minutes until a few bubbles form at the edges or on the surface. Flip the cakes and cook until lightly colored on the bottom and cooked through in the center, about 30 seconds. Repeat with the remaining batter, adding butter to the skillet as needed. Serve immediately or keep hot in a 200°F oven, loosely covered with foil, for up to 20 minutes. Leftover batter keeps in the refrigerator, covered, for at least 1 day.

BASIC WAFFLES

MAKES ABOUT FIVE 7-INCH WAFFLES

1¼ cups flour: white or brown rice, gluten-free oat, corn, buckwheat, chestnut, teff, sorghum, or dried shredded coconut, or nut flour/meal (see pages 353–354 for equivalent weights for each flour)

If using dried shredded coconut or a nut flour other than chestnut, add ½ cup plus 2 tablespoons (95 grams) white rice flour

2 teaspoons baking powder

Slightly rounded ½ teaspoon salt

2 large eggs, separated, at room temperature

6 tablespoons (85 grams) unsalted butter, melted and warm, plus more for the waffle iron

1 cup milk, warm

⅛ teaspoon cream of tartar

1 tablespoon sugar

Whisk the flour(s), or coconut or nut flour and rice flour, with the baking powder and salt in a medium bowl. Add the egg yolks, butter (use the greater amount for richer and crispier waffles), and milk and whisk until smooth and well blended.

In a clean, dry bowl, beat the egg whites and cream of tartar until the egg whites are creamy white and hold a soft shape when the beaters are lifted. Sprinkle in the sugar and beat until stiff but not dry. Fold the egg whites into the batter.

Heat a waffle iron and brush it with butter. Cook the waffles according to the manufacturer's directions. Serve immediately or keep hot in a 200°F oven, loosely covered with foil, for up to 20 minutes. Or let cool completely and reheat in a toaster or toaster oven. Leftover waffles keep better than leftover batter; store them in an airtight container in the freezer. Reheat in a toaster or toaster oven.

BASIC CREPES

MAKES 18 OR MORE 6-INCH CREPES

1¼ cups flour: white or brown rice, gluten-free oat, corn, buckwheat, chestnut, teff, or sorghum (see pages 353–354 for equivalent weights for each flour)

2 tablespoons (30 grams) unsalted butter, melted, plus more for the pan

3 large eggs

1 tablespoon sugar (optional)

Scant ¼ teaspoon salt

1½ cups whole milk, plus more as needed

¼ cup water

In a medium bowl, whisk the flour with the butter, eggs, sugar (if using), salt, and about one-quarter of the milk in a medium bowl until blended and very smooth. Whisk in the remaining milk and the water. Cover and refrigerate the batter for at least an hour (to let the flour hydrate) or up to 2 days. Stir the batter well before and frequently as you use it.

Heat a 6-inch crepe pan or skillet over medium-high heat. Brush the pan lightly with butter. When a drop of water sizzles on the pan, lift the pan off the burner and pour in 2 tablespoons of the batter. Immediately tilt and rotate the pan, shaking as necessary to coat the bottom of the pan entirely. Fill in any holes with extra drops of batter. Set the pan back on the burner and cook until the surface of the crepe no longer looks wet and the underside is golden brown, 30 to 60 seconds. If the crepes will be filled and sautéed before serving (as for blintzes), cook on one side only, then loosen the edges of the crepe with a spatula and invert it onto a piece of wax paper or a paper towel. Otherwise, turn the crepe and cook the other side for 10 to 20 seconds longer, then slide or flip it onto the paper. Repeat with the remaining batter, buttering the pan only as necessary. If the crepes seem too thick, adjust the batter with a little extra milk.

Crepes can be stacked as they come out of the pan—they won't stick to each other so there is no need to put wax paper between them, unless you are freezing them and plan to defrost only a portion at a time; then you should put a double sheet of wax paper between portions. When the stack is cool, it can be wrapped airtight and stored in the refrigerator for up to 2 days, or up to 3 months in the freezer. Defrost crepes in the refrigerator or in a covered dish in a 300°F oven for about 10 minutes. For serving suggestions, see opposite.

Garnishes for Plated Crepes

You can make filled or folded and garnished crepes or crepe cakes from the basic recipe. Coordinate the filling and garnishes to complement or contrast with the flavor of the crepes.

White Rice Flour Crepes

- Sautéed pineapple fans and coconut ice cream
- Add ½ teaspoon ground cinnamon to the batter; top with chocolate ice cream.

Brown Rice Flour Crepes

- Lemon Curd (page 341) with sliced strawberries
- Coffee ice cream with toasted walnuts or slivered dates, or Prunes Poached in Coffee and Brandy (page 336)

Oat Flour Crepes

- Vanilla ice cream with toasted pecans and sliced bananas
- Vanilla ice cream and salted caramel sauce
- Strawberries with brown sugar or dark muscovado sugar
- Dulce de leche ice cream, blackberries, and toasted walnuts

Corn Flour Crepes

- Vanilla ice cream and dulce de leche or cajeta sauce and chopped toasted pecans
- Crumbled queso fresco with a drizzle of honey

Buckwheat Flour Crepes

- Sour cream and/or a drizzle of honey
- Fresh blackberries and a dusting of powdered sugar
- Vanilla ice cream and sautéed cherries
- Crème fraîche or sour cream and cherry preserves

Chestnut Flour Crepes

- Chopped marrons glacés (or chestnuts in syrup) and vanilla ice cream
- Sliced oranges with orange flour water, whipped cream, and a drizzle of honey
- Vanilla ice cream and sautéed apples or pears or spicy apple or pear butter
- Vanilla or coffee ice cream and chopped Walnut Praline Brittle (page 337)

Teff Crepes (or Cocoa Crepes, page 253)

- Chocolate ice cream and chopped Walnut Praline Brittle (page 337)
- Folded with Nutella, topped with vanilla ice cream and chocolate sauce
- Folded with apricot preserves, topped with vanilla ice cream and chopped toasted hazelnuts

Sorghum Flour Crepes

- Peach ice cream with chopped candied ginger
- Vanilla ice cream with salted peanuts and sorghum syrup
- Toasted pecans, bananas, and whipped cream
- Vanilla ice cream with rum-soaked raisins and freshly grated nutmeg

TIPS FOR MAKING CREPES

Here's how to figure out how much batter to use to make a crepe, depending on the size of your pan and how many crepes you might be able to make per batch once you get the hang of it.

Find your pan's true size by measuring across the flat surface (the bottom) of the pan. The crepe recipes in this book make about 2¾ cups (44 tablespoons) of batter; the more you perfect your technique, the more crepes you will get from each batch.

- For a 6-inch pan: use 2 tablespoons of batter per crepe to make up to 22 crepes per recipe
- For an 8-inch pan: use 4 tablespoons of batter per crepe to make up to 11 crepes per recipe
- For a 10-inch pan: use 6 tablespoons of batter per crepe to make up to 7 crepes per recipe

It's easier to make crepes in a small pan than a large one. Start small if you are a newbie. Some people find a nonstick pan easier. Either way, the pan should be perfectly smooth and clean.

- Use melted butter or clarified butter to grease the pan lightly but thoroughly. Recoat the pan from time to time, but only as necessary.
- Before you pour the batter, flick a drop of water into the pan; it should sizzle. If the pan is not hot enough, the crepe will stick.
- Learn to pour the batter and tilt the pan so that it is completely coated, without using excessive amounts of batter. You can also skip this challenge by purposely pouring too much batter into the pan and pouring any excess back into the batter bowl as soon as the pan is coated. Either way, expect some thicker crepes or some goof-ups before you get it right.
- Lift the pan off the burner before you pour the batter into it. Use a small ladle or measuring cup and hold it close to the pan when you pour.
- Pour the batter into the middle of the pan, then tilt and rotate it—not too quickly—and shake it as necessary so that the batter spirals toward the edge of the pan without crossing back and forth over areas that are already coated. Fill in gaps with drops of extra batter from the ladle. It doesn't have to be perfect.
- Put the pan back on the heat after it's covered with batter.
- The cooking time is brief, only 30 to 60 seconds. The surface of the crepe should no longer look wet and the underside should be golden brown. But crepes are forgiving; a few seconds more or less won't hurt.
- Use a spatula to loosen the edges of the crepe—if the crepe is not adequately cooked, it will not loosen. Once the edge can be loosened, use the spatula or even your fingers to lift and flip the crepe. Some batters are more fragile than others, but if the crepe is too fragile to lift, it may need to cook for a few more seconds.
- If the batter or crepes are too thick, you can thin the batter with extra milk or water.

BUYING AND STORING FLOURS

Flavor flours often have a shorter shelf life than wheat-based flours, so buy them from sources that have high turnover, and always look at "sell by" and "best by" dates on packages.

All of the flours should be stored in airtight containers away from light and heat. The whole-grain flours (oat, sorghum, corn, brown rice, teff, and buckwheat and nut flours—all of the flours other than white rice and coconut flour) will eventually go rancid at room temperature. Unopened packages can be kept in a cool, dark pantry at least until their "sell by" dates. Once opened, flours should be stored in airtight containers with tight lids or in heavy resealable zipper-lock bags. If you buy flours from Authentic Foods (see Resources, page 351), you can keep the flour in the tough, square-bottomed bags they come in: like canisters, these are easy to scoop from and they don't topple over on the shelf. (Their new zipper closures make the opening perfect for pouring but too small to scoop from; I cut them off, fold down the top edges of the bag, and secure them with a large binder clip or wide tape to keep the contents airtight.) If you buy flours in cellophane bags, consider transferring the bags or just the contents to airtight plastic containers or large Mason jars; these are easier to scoop from and neater on the shelf. Flours should keep this way for at least 2 to 3 months in a cool, dry place and at least 6 months or more in the refrigerator or freezer; the less air in the container the better. (Measure or weigh the amount of flour needed for a recipe while the flour is still frozen: to avoid repeated defrosting and condensation in the container, cover and return it to the freezer immediately.) Buy conservative amounts of flour and label them with your purchase date and/or the "sell by" date when you transfer them to new containers. Rotate flour in canisters when you replenish your supply: dump any flour remaining in the bottom onto waxed paper, pour in the new supply, and pour the old flour on top. Wash and dry the container from time to time as well. Always smell the flour before using it; it should smell sweet and fresh, never sour or rancid.

SHOPPING GLUTEN FREE

If you have celiac disease or are highly sensitive to gluten, you are probably already a close reader of ingredient labels. You probably also know not to buy gluten-free flours (or anything) from bulk bins to avoid inadvertent contamination. Packaged flours that are labeled "gluten free" must meet FDA standards. In the case of oats and oat flour, it means that they were processed and packaged in facilities that are gluten free, and were not grown in fields adjacent to wheat. Products that are "certified" gluten free meet even higher standards than those required by the FDA. Read more about FDA regulation and gluten-free certification online (see Resources, page 351).

THE FINENESS OF FLOUR: SIZE MATTERS

The same type of flour from different sources may differ in particle size: some rice flours are fine, some "superfine," and some are finer still. The fineness of the flour affects the texture of your results and complicates measuring. I compared rice flour from Bob's Red Mill, from Authentic Foods (which is labeled "superfine"), and a Thai brand of regular rice flour (not sweet rice flour) from an Asian grocery store (which is even finer). The finer flours from Thailand and Authentic Foods often produced the best results, but all of the flours produced good results in most of the recipes. Puddings and pastry cream proved one type of exception. You will not get the "silky" results promised in these recipes unless you use one of the finer flours. Meanwhile, the very finest (Thai) flour produced chiffon cakes and sponge cakes that were too dense. Of course, each recipe specifies which flour to use where it is critical.

Although volume measures are always given, you will get better results with flours that vary in fineness—at least for the recipes in *this* book—if you measure by weight instead of volume. Authentic Foods superfine white rice flour and Bob's Red Mill white rice flour, which is less fine, weigh between 140 grams and 160 grams per cup. But Thai rice flour is extremely fine and weighs only about 100 grams per cup. There are differences among glutinous rice flours, too. Two volume measurements are given for rice flour: one for Authentic or Bob's Red Mill and another for Thai flour. If you use another flour and do not know how it compares to any of these, you would be wise to measure by weight, using a scale (see "Why Weigh?," below).

WHY WEIGH?

In all baking, a scale is the simplest and easiest way to be sure that you are using the same amount of the ingredient as the person who created the recipes. Flour is one of the trickiest of all ingredients to measure. For one thing, measuring styles vary. A cup of any single flour can vary in weight by as much as 50 percent depending on how the cook used the cup: Did she dip the measuring cup into a compacted bag of flour and press it against the side of the bag to level it (*please, please* never measure this way!), or dip it into a canister of loosened flour, or spoon the flour lightly into the cup and then sweep it level? Each method affects how much flour ends up in the cup. In addition, flour that is compacted in a bag will weigh more per cup than flour that has been stirred of fluffed or even just handled before measuring. But if we all use a scale, your 100 grams of flour is the same quantity as my 100 grams of flour.

To make matters even more interesting, weight and volume are not the same for all flours. A cup of corn flour does not weigh the same as a cup of rice flour or a cup of chestnut flour. Even more startling, there is no standard fineness for the same flour from different mills (see "The Fineness of Flour: Size Matters," opposite): so even the same type of flour may not weigh the same per cup from one brand to another. The only way to deal with this chaos is to measure by weight.

Fortunately, once you get the hang of it, a scale is the simplest and cleanest way to measure. Fewer utensils are required, so there is less cleanup. There are other advantages, too.

Say a recipe calls for 100 grams of chopped nuts or ground nuts; all you have to do is weigh the nuts first (no matter what size they are to start with), then chop or grind them, instead of chopping or grinding a little at a time until your measuring cup is filled, and then usually having some extra left over. With a scale you can often weigh ingredients one by one, directly into the mixing bowl; just press the tare to reset the scale to zero after each ingredient is added before weighing the next. Blissfully easy!

All this being said, I have provided cup measurements as well as weights for all of the recipes, and in the case of rice flour, I've given different cup measurements for the different types. But I still hope you will get and use a scale.

A battery-powered digital scale is most efficient; choose one that will switch between ounces (imperial) and grams (metric), ideally with increments of 5 (or fewer) grams or one-tenth of an ounce (see Resources, page 351).

How to Measure Flour Using Cups

The best way to measure flour, or any dry ingredient, for recipes is to use a scale (see above). That being said, here is how to use measuring cups to get as close as possible to the listed flour (or cocoa powder) weights in each recipe.

Use dry measures (see page 42). Set the appropriate size measuring cup on a piece of wax paper on the counter. If the flour in the canister or the bag is very compacted, loosen it gently with a spoon (don't overdo it, though, or your measure will be too light). Pour or spoon the flour lightly into the cup until it is heaped above the rim. Without tapping or shaking the cup or compacting the flour, sweep the flour level with the rim using a straight edge.

OVEN RACK POSITION

Oven rack position affects performance, including baking time, degree of browning on top or bottom, and sometimes even whether or not a cake will rise properly. Most recipes call for positioning a rack in the lower third of the oven; this means the rack should be placed a third of the way up from the bottom of the oven or just below the center, so that the actual cake pan is almost in the center of the oven rather than in the upper half, which might

cause a cake to brown too much or too fast before the interior is done. A single sheet of cookies or a very thin cake such as a sponge sheet may be placed on a rack in the center of the oven, but two sheets should go in the upper and lower thirds and should be rotated from upper to lower (and front to back) halfway through the baking time. Items that need a good hit of heat on the bottom of the pan—think upside-down cakes with gooey sweet fruit on the bottom—may be placed on the bottom rack. I like to position biscuits and scones in the upper third so they brown nicely. All this is to say that the different oven positions are given for each recipe, and each has a rationale.

WHY AND HOW TO LINE PANS

Recipes frequently call for lining pans, either with parchment paper or foil depending on the recipe. Pan liners not only make removal of the cake or cookies from the pan simpler, they can also eliminate cleanup, and even improve results by protecting the cakes or cookies from overbrowning if pans are too dark or too thin. It also has to be said that loaf cakes baked in pans lined fully—bottom and all four sides—with parchment paper always seem to look both handcrafted *and* professional.

It's easy to line a baking sheet—just plop the liner on the pan and be done with it. Slip parchment circles or squares into the bottom of cake pans; there is no need to grease the parchment or the pan under it. For brownies, you can also use the time-honored method of lining pans across the bottom and up two opposite ends with a little excess to use as a handle. But it takes no more time and effort to line the bottom and all four sides of brownie pans and loaf pans if you know the trick: Use a piece of parchment or foil (foil is easier but not as attractive for loaf cakes) about 4 inches longer and wider than the bottom of the brownie pan or 5 inches longer and wider than the loaf pan. Turn the pan upside down on the counter and center the liner on it with the excess extending 2 to 2½ inches on all sides. Fold the sides and ends of the liner down against the pan sides and crease the folds. Fold the corner "wings" over the ends of the pan and crease the folds. Slip the creased liner off the pan and fit it inside the pan, adjusting the folds and creases as necessary.

INGREDIENTS

Except for the flours themselves and perhaps xanthan gum, this list will look familiar to anyone who bakes, even occasionally. In other words, there is no need to create an entirely new pantry to bake with flavor flours. The information on different types of chocolate and cocoa should be useful to anyone who needs to brush up on the newer chocolates, percentage labeling, or different types of cocoa.

Baking Powder and Baking Soda

Non-aluminum gluten-free baking powder was used for testing all the recipes in this book. Baking powder loses potency with time; watch the expiration date and keep the container covered with its snap-on lid. If you bake infrequently or suspect dead powder, add about 1 teaspoon of it to a cup of hot water. If it bubbles vigorously, use it. If not, toss it.

Baking soda appears to last indefinitely, even when the container is not sealed, but if in doubt, spoon a little into a cup and add vinegar. It should bubble vigorously.

Buckwheat Flour

See page 168.

Butter

Recipes in this book were tested with regular unsalted butter—not European or European-style butter. If you use salted butter, subtract at least ¼ teaspoon salt from the salt in the recipe for each 8 tablespoons (113 grams) of butter used. If you must use margarine or a nondairy spread, choose one labeled suitable for baking (with about 11 grams of fat per tablespoon), as some vegetable spreads and butter substitutes contain too much water (and not enough fat) to substitute directly for butter. Do not substitute liquid fats or oils for butter.

Cajeta

See Dulce de Leche, page 38.

Chestnut Flour

See page 198.

Chocolate

CHOCOLATE CHIPS AND CHUNKS

Purchased chocolate chips (and some chocolate chunks) are specially formulated with less cocoa butter than bar chocolate so that they hold their shape when baked into cookies. They may also help keep cookies from flattening out during baking. Since chocolate chips (and similarly formulated chunks) stay thick when melted and are usually fairly sweet, I don't recommend them for melting and blending into batters unless specifically called for. By contrast, you can chop your favorite chocolate bars instead of using purchased chocolate chips or chunks, as long as you don't mind cookies that are a little flatter and chunks that flow a little rather than hold their original shape.

COCOA POWDER

Unsweetened cocoa powder is made by removing 75 to 85 percent of the fat (cocoa butter) from chocolate liquor, then pulverizing the partially defatted substance that remains.

The result is pure, natural (nonalkalized) cocoa powder. Dutch-process (alkalized) cocoa is processed with an alkali to reduce acidity and harshness. Alkalizing also darkens the color of the cocoa and imparts a flavor best described as "Oreo cookie." Chefs and consumers have different preferences when it comes to natural or Dutch-process cocoa. Regardless of which you prefer, the two types are not always interchangeable in recipes. Always use the type of cocoa called for, unless a choice is given.

MILK CHOCOLATE

Milk chocolate is sweetened chocolate that must contain at least 10 percent cocoa beans (cacao) and 12 percent milk solids, plus milk fat. Most milk chocolates today exceed the minimum requirements for cocoa beans considerably, resulting in chocolate with more chocolate flavor and less sweetness. The recipes in this book were tested with milk chocolate with at least 33 percent cocoa beans (cacao).

ROASTED CACAO NIBS

Cacao nibs (aka cocoa nibs) are pieces of hulled cocoa beans—the essential ingredient in all chocolate. Nibs are crunchy and somewhat bitter. Crushed, ground, or left whole, they add unique chocolate flavor to all kinds of desserts. Raw nibs are available, but I prefer the flavor of roasted nibs.

SEMISWEET, BITTERSWEET, AND DARK CHOCOLATE

These are sweetened dark chocolates: pure ground cocoa beans, optional extra cocoa butter, sugar, optional lecithin and/or vanilla, and sometimes a small amount of milk have been added. The standard brands in the baking aisle contain 50 to 60 percent cacao and 40 to 50 percent sugar. But there are dozens of semisweet and bittersweet chocolates (some simply labeled "dark") with much higher cacao percentages. And while it is reasonable to presume that bittersweet is less sweet than semisweet, there is no official distinction between them; so one brand of bittersweet may be sweeter than another brand of semisweet. Cacao percentage is a better predictor of sweetness, chocolate intensity, and behavior in recipes than any of the terms. Recipes in this book call for dark chocolate by percentage, or give a range of percentages, to ensure that you have the best possible results.

UNSWEETENED CHOCOLATE

Technically called chocolate liquor (though it contains no alcohol), unsweetened chocolate is pure ground cacao nibs, often with a fraction of a percent of lecithin as an emulsifier and sometimes a bit of vanilla. Unsweetened chocolate may be labeled 99 percent cacao (to account for that tiny amount of lecithin and/or vanilla) or 100 percent cacao. Though very strong and bitter, the highest quality unsweetened chocolate is smooth enough and palatable enough to nibble.

HOW TO MELT CHOCOLATE

Most cookbooks advise melting chocolate in a double boiler or in a bowl suspended over a pot of simmering water. But I think an open water bath is more flexible, easier to control, and safer.

Either way, the object is to control the heat enough to prevent overheating the chocolate. The water in a double boiler is hidden from view in a closed chamber, where it is extremely likely to come to a full boil without your noticing, especially if you are lulled to inattention by the implied safety of a double boiler. Steam in a closed chamber is far hotter than boiling water, and your bowl of chocolate is sitting in that steam. Moreover, the common technique of creating a makeshift double boiler by setting a bowl over a pot of water carries additional risk when the bowl is much wider than the pot: the sides and rim of the bowl (and any chocolate in the bowl) above the rim of the pot are exposed to scorching heat coming up the sides of the pot from the burner beneath it.

An open water bath has the advantage of complete visibility; it is hard not to notice when the water begins to simmer or boil (the bowl often rattles as well), and then you can turn the heat down or even off. Although the bowl of chocolate sits on the bottom of the skillet, there is actually a tiny film of water between the bowl and the pan; this layer buffers some of the heat. And the chocolate around the sides of the bowl doesn't get scorched, because the bowl doesn't extend beyond the sides of the bath.

Into the bargain, the bath allows you to melt the chocolate in an appropriately sized bowl: a small bowl for a few ounces or a large bowl if the recipe calls for adding the rest of the ingredients into the chocolate later.

If you still prefer your double boiler, don't ever worry again if the container of chocolate touches the water beneath it—it may even be safer if it does!

Before You Start

Chocolate should always be chopped before melting so that it melts easily with gentle heat. Dark chocolate (including semisweet, bittersweet, and unsweetened) may be chopped coarsely. White and milk chocolate should be finely chopped.

The cutting board, bowl, and all utensils should be dry—take care not to let any splash or drip of liquid come in contact with the chocolate. These are best practices for handling chocolate safely, though they are truly critical only when chocolate is melted solo, with no added fat or liquid—then it is vulnerable to any contact with small amounts of moisture or liquid, which might cause the chocolate to thicken or seize, instead of melt smoothly.

How to Melt Chocolate in a Water Bath

Put the chopped chocolate in a stainless steel bowl. For dark chocolate, set the bowl directly in a skillet of barely simmering (or not even simmering) water. The bowl of chocolate does indeed touch the water and the bottom of the skillet. The skillet should be several inches wider than the bowl, to prevent the edges of the bowl from getting too hot to touch. (See the photo on page 36.) Stir the chocolate frequently until it is almost entirely melted—or as directed in the recipe.

For white or milk chocolate, which burn very easily, remove the skillet of simmering water from the heat and wait 60 seconds before setting the bowl of chocolate in the water. Then stir the chocolate, letting the hot water do the job without any live heat under the pan.

How to Melt Chocolate in the Microwave

Chop dark chocolates (including semisweet, bittersweet, and unsweetened) coarsely into pieces about the size of almonds; chop milk chocolate or white chocolate very fine. Put the chocolate in a perfectly dry microwave-safe container. Heat dark chocolate on Medium (50 percent) power, and milk and white chocolate on Low (30 percent or Defrost). Start with 1 to 2 minutes for amounts up to 3 ounces, and 3 minutes for larger amounts. Even if most of the chocolate is unmelted, stir it well before microwaving for additional increments of 5 to 15 seconds or more, depending on how much of the chocolate is left unmelted after stirring well. Be conservative; the goal is warm chocolate, not hot chocolate.

WHITE CHOCOLATE

White chocolate is made from only the fat (cocoa butter) of the cocoa bean, rather than the whole cocoa bean, and is combined with sugar, dry milk solids, milk fat, lecithin, and vanilla. White chocolate is now recognized and defined by the FDA as a form of real chocolate to distinguish it from "white confectionery coating," which is made from tropical fats other than cocoa butter, and thus contains not a single ingredient derived from cocoa beans.

Coconut, Dried—Shredded or Flaked

With the exception of German Chocolate Cake (page 237), which calls for sweetened coconut, recipes in this book call for unsweetened shredded coconut and/or wider shavings of unsweetened coconut (aka coconut chips or flakes). Both may be found in the baking aisle or the bulk food aisle of better supermarkets and natural food stores.

Coconut Flour

See page 296.

Coffee and Instant Espresso Powder

If a recipe calls for freshly ground coffee beans or brewed coffee, I use beans from a specialty purveyor, rather than a vacuum-packed brand, and I grind them myself. For recipes that call for instant espresso powder, use an unsweetened brand that does not contain milk, such as Medaglio d'Oro or King Arthur; or use regular powdered coffee or coffee crystals but increase the amount by about 25 percent.

Corn Flour

See page 138.

Cornmeal, Stone Ground

Stone-ground cornmeal is made from corn that has not been degerminated. It is whole-grain flour and should be stored in the refrigerator or freezer to prevent rancidity.

Cream

Buy heavy cream or heavy whipping cream (with at least 5.5 grams of fat per 2-tablespoon serving) rather than whipping cream if you have a choice, because it whips up more reliably. The best-tasting cream is simply pasteurized, not ultra-pasteurized or "sterilized" (for longer shelf life), and it contains a single ingredient: cream. If you need to keep cream on hand for a spur-of-the-moment need, keep a carton of ultra-pasteurized in the back of the fridge—it will last for quite a while unopened. When you know you are going to use it to make dessert, buy the good stuff, if you can find it.

Dried Fruit

Dried fruit should be moist and flavorful, rather than hard and dry. Whole pieces are fresher, plumper, and more flavorful than prechopped or extruded pellets. Chop your own, using an oiled knife or scissors.

Dulce de Leche

Dulce de leche and/or cajeta may be found in better supermarkets, and in Latin American or Mexican groceries, as well as online. Both are caramelized milk sauces, with cajeta traditionally made with goat's milk or half goat's and half cow's milk. The terms are used interchangeably, so you must read the labels if you are looking for one milk or the other.

Ghee

Ghee may be substituted for clarified butter in any recipe in this book. Ghee is a type of clarified butter—pure butter oil minus the water and milk solids found in regular butter—used widely in Indian and other South Asian cooking. Ghee has a slightly nutty flavor, even more delicious than that of clarified butter, because the milk solids are cooked and caramelized before they are removed. The shelf life of ghee is longer than that of clarified butter because the latter usually has some traces of milk left in it that will go sour unless you keep it in the freezer. Ghee is available in better supermarkets and Indian or South Asian groceries.

Nuts

For maximum freshness and flavor, buy nuts raw and toast them yourself. If possible, buy nuts in bulk from a store with high turnover, and taste before buying. Big-box stores are also a good source. Whole nuts (almonds and hazelnuts) or halves and large pieces (walnuts and pecans) stay fresher longer; it's better to chop them yourself.

Nut Flour(s)

See page 296.

Oat Flour

See page 84.

Potato Starch

Not to be confused with potato flour (which is made from the entire potato), potato starch—also called potato starch flour, just to confuse things—is derived only from the starchy part of the potato. It has a relatively neutral flavor and a light silky texture that, combined with oat flour and brown rice flour, produces superb Chocolate Chip Cookies (page 126) and New Classic Blondies (page 118).

Rice Flour, Brown

See page 52.

Rice Flour, Sweet or Glutinous

See page 52.

Rice Flour, White (also Thai Rice Flour)

See page 52.

Salt

The recipes in this book were tested with fine sea salt, but regular table salt may be used.

Sorghum Flour

See page 262.

Spices

Ground spices should still smell potent in the jar; don't expect them to keep forever. Keep whole nutmeg, cinnamon sticks, and cardamom pods on hand, in addition to the ground spices, for special needs. Use a Microplane zester to grate a little nutmeg or some cinnamon stick over brownies or cookies or over hot doughnuts just before serving. To use cardamom seeds from whole pods: Crush the pods lightly with a heavy object to split them. Pick off and discard the papery dry pod to find the black or gray seeds. Crush the seeds or pulverize them in a mortar, as directed in the recipe.

Sugar

BROWN SUGAR

Brown sugars were once semirefined sugars with some of the natural molasses left in them. Commercial brown sugar today is retrofitted: it is refined white sugar with varying amounts of molasses added. Brown sugar imparts delicate butterscotch flavors to baked goods and desserts. Recipes may specify a preference for light or dark brown sugar, but normally you can use them interchangeably. Brown sugar hardens with exposure to the air; store it in an airtight container or tightly sealed in the bag it came in. The sugar should be lump-free before it is added to a batter or dough, since it is unlikely to smooth out; squeeze lumps with your fingers or mash them with a fork. To soften hardened brown sugar, sprinkle it with a little water, put it in a tightly covered container (or wrap tightly in foil), and place in a 250°F oven for a few minutes. Cool without unwrapping before using. Measure brown sugar using a scale, if possible, or by packing it firmly into a measuring cup.

GRANULATED WHITE SUGAR

The recipes in this book were tested with C&H granulated cane sugar (a brand sold mainly west of the Mississippi). Sugars vary in different parts of the country. If your brand of sugar is coarser than regular table salt, and/or you think that your cakes or cookies could be more tender, use baker's sugar, or superfine or bar sugar, or process your granulated sugar briefly in the food processor before using it. Beet sugar is chemically identical to cane sugar, but many bakers have reported differences and disappointments with beet sugar. Stick with cane sugar.

POWDERED SUGAR

Also called confectioners' sugar or icing sugar, this is granulated sugar that has been pulverized and mixed with a little cornstarch to prevent clumping. We use powdered sugar mostly for dusting lightly over desserts to soften or dress up the look.

Teff Flour

See page 230.

Vanilla

Pure vanilla extract is a must—it simply tastes better than imitation. I use bourbon (aka Madagascar) vanilla for most things where an expected vanilla flavor is wanted, but I enjoy having Mexican and Tahitian vanillas on hand for special projects. See Resources (page 351) for a supplier of great vanilla.

Xanthan Gum

Xanthan gum is produced through the fermentation of a bacterium called *Xanthomonas campestris* on corn syrup or on other mediums such as wheat or soy. People with wheat allergies should purchase xanthan gum from a reliable gluten-free source. Xanthan gum is used in commercial food products to add volume, thicken and stabilize liquids, and hold particles in suspension (think salad dressing). It is also used to create a smooth texture in commercial ice creams and frozen desserts. And in gluten-free baking, it is used to bind ingredients and/or create chewiness in the absence of gluten. Powdered xanthan gum is available in the baking aisles of better supermarkets or online. See Resources, page 351.

EQUIPMENT

If you bake, you may already have what you need to make all of the recipes in this book. For those who are still acquiring the tools of the trade, here's an annotated inventory of a good baker's kitchen.

Baking Dishes and Pie Plates (Glass or Ceramic)

Fruit desserts in this book call for 2-quart baking dishes in any shape, 2 to 3 inches deep, and pies call for 9-inch pie plates.

Baking Mat—Silicone

For thin, crispy cookies like tuiles, or any cookies meant to be uniformly crispy, a silicone mat is a good option (though not essential) for lining baking sheets. However, silicone is not the best liner for most other cookies, and I don't consider it a must-have item.

Baking Pans (Metal) for Cakes, Brownies, and Loaves

The recipes in this book call for 8-inch and 9-inch round and square pans, 2 inches deep; 8- and 9-inch springform pans or cheesecake pans with removable bottoms, 2½ to 3 inches deep; 4-cup and 6-cup loaf pans; 10-inch (10- to 12-cup) tube pans with removable bottoms; and 10- to 12-cup Bundt pans.

For even baking and moist, tender cake layers, medium-weight to heavy light-colored aluminum pans generally work best. Metal pans that are thin or dark produce cakes that overbake at the sides and on the bottom, sometimes even before the inside is done. (Glass pans may do the same—don't use them unless they are called for.) There are some exceptions: some pound cakes and rich cakes baked in heavy (often decorative and often dark) tube, Bundt, or loaf pans acquire a deep golden-brown crust that is both delicious and beautiful.

Baking Sheets

Medium-weight to heavy light-colored aluminum baking sheets and jelly roll pans will not warp or bend and they cook evenly, without hot spots. Avoid dark or nonstick surfaces, which can overbake and toughen tender sheet cakes and cookies. For both cookies and sheet cakes or roulade sheets, I like medium-weight commercial "half sheet pans," which measure 12 by 16 inches by 1 inch—nothing fancier. If cookies are baking unevenly or browning too quickly on the bottom and edges—before the tops are baked—your pans may be too thin or too dark. In a pinch, parchment liners may help promote even baking, even if your pans are less than ideal.

Bowls

A baker's kitchen needs a variety of bowls of different sizes. They need not be expensive. Glass and stainless steel bowls are good for different tasks. Glass is great for use in the microwave or when you need the added stability of its weight—for instance, when you must whisk or beat with one hand and pour in an ingredient with the other. But stainless steel is more versatile. It is preferable for melting chocolate or heating anything in a water bath or improvised double boiler. Bowls that are nearly as tall as they are wide are best for beating egg whites and for keeping sugar and flour from flying out when using a handheld mixer.

Cooling Racks

Racks allow air to circulate under cakes and cookies while cooling.

Deep-Fat Fryer

Doughnut makers find that a fryer, even a small one, with a temperature gauge produces better (less greasy) doughnuts without breaking down cooking oils.

Food Processor/Blender/Coffee Grinder

A food processor pulverizes nuts (with the steel blade) or grates them (with the fine shredding disk) to make nut flours. It also mixes cookie dough and cake batter in a flash. A blender is good for liquid ingredients or purees, or for turning nuts or buckwheat groats into meal, but is apt to turn large quantities of nuts into paste rather than meal or flour. A coffee grinder is best for making flaxseed meal with up to ⅓ cup (about 50 grams) of flaxseeds; for up to ⅔ cup (100 grams), use a blender instead.

Measures: Dry and Liquid

Dry measuring cups are designed to measure dry ingredients such as flour. Metal or plastic dry measures come in sets of ¼, ⅓, ½, and 1 cup. These measures are meant to be filled

Tube Pan Tricks

In the traditional baking repertoire (that is, baking with wheat flour), light, airy cakes with delicate structures, like angel food, chiffon, and certain light sponge cakes, are baked in large, tall tube pans. The tube not only helps these cakes bake evenly both inside and outside, it also prevents delicate cakes from sinking in the center as they cool, which they might otherwise do. Decorative tube pans, such as Bundt pans, are also used for some heavier and sweeter cakes (think rich coffee cake) that would ordinarily sink in the center.

Since the flours used in this book do not contain gluten, tube pans provide essential support in several recipes. Sponge and chiffon cakes stick to the contours of decorative Bundt pans, but they do very well in plain tube pans with removable bottoms—the type traditionally used for angel food cakes. Rich butter cakes do well in plain tube pans or Bundt pans. Each recipe will specify the right type of pan.

level with the rim: use a ½-cup measure to measure ½ cup; don't shake what you think is ½ cup into a 1-cup measure. Liquid measures are clear glass or plastic pitchers with measures marked up the sides: pour liquid ingredients to the appropriate mark with your head lowered so you can see the amount at eye level.

Measuring Spoons

Metal or plastic measuring spoons come in sets of ¼ (1.25 ml), ½ (2.5 ml), and 1 teaspoon (5 ml), plus 1 tablespoon (15 ml, which equals 3 teaspoons). Some sets include ⅛ teaspoon. Measures are meant to be leveled. I don't bother to use measuring spoons that purport to measure a pinch or a smidgen.

Nut Grinder

A nut grinder allows you to make nut flours that are fresher and more flavorful than store-bought. The best grinders produce fine, dry (not oily) flours by allowing the nuts to pass through a grating or shredding device rather than keeping them in continuous contact with a blade. I use an inexpensive Swedish grinder with a barrel-shaped grater/shredder and hand crank; it clamps to a cutting board or tabletop and has lasted for decades.

Microwave Oven

A microwave oven is good for reheating sauces; softening, melting, or even clarifying butter (see page 335); liquefying crystallized honey; defrosting ingredients; and softening rock-hard ice cream to scooping consistency. A microwave can also be used (very, very carefully) to bring cold dairy products—milk, eggs, and cheese—to room temperature.

Mixers

Recipes that call for an electric mixer most often call for a stand mixer with a whisk or paddle attachment, depending on the recipe. Batters and doughs without gluten often get structure from whipped eggs or egg whites; a stand mixer with a whisk attachment always outperforms a handheld mixer and usually in less than half the time. The paddle attachment similarly handles very thick, sticky doughs. A few recipes call for a handheld mixer.

Parchment and Wax Paper

Parchment paper makes a superb and reliable pan liner for cake pans or baking sheets when a liner is called for. Wax paper is less expensive than parchment and is a good surface for measuring and/or sifting flour; layering between cookies, crepes, or cake layers; or performing miscellaneous tasks that do not involve the oven or high temperatures.

Pastry Bags and Tips

Unlike cloth pastry bags, nylon or polypropylene wash easily and don't become sour smelling, nor does the moisture from whipped cream weep through them. Disposable plastic bags may even be washed and reused. Bags should be bigger than you think you need and never filled more than half full: a 16- to 18-inch bag is good for piping a batch of meringues or lots of whipped cream. A 10- to 12-inch bag is good for smaller quantities.

Pastry Brushes

Natural boar bristle brushes are a pain in the neck to wash, but they do a better job than nylon or other synthetic bristles—including silicone–for baking tasks where a thin but thorough coating of something (egg wash, melted butter, etc.) is needed, or even for moistening cake layers. Keep at least one natural bristle brush dedicated to desserts so you won't have to worry about the lingering flavor of barbecue sauce.

Scale

A scale is the best and simplest way to measure almost everything, especially dry ingredients like flour and cocoa powder. A battery-powered digital scale is most efficient; choose one that will switch between ounces (imperial) and grams (metric), ideally with increments of 5 (or fewer) grams or one-tenth of an ounce.

Serrated Bread Knife

A serrated knife with a 12-inch blade is ideal for cutting cakes into thin horizontal layers, but in a pinch, a shorter knife will do the job.

Skewers and Toothpicks

For testing cakes, wooden toothpicks or slim bamboo skewers are better than metal "cake testers." Moist batter and crumbs stick to wood much better than to metal, so you can see exactly how gooey or dry your cake is within.

Spatulas—Metal

A metal spatula with a straight 8-inch blade is perfect for frosting cakes. An offset spatula (with a bend in it right after the handle) at least 8 inches long is good for spreading a thin, even layer of batter into a jelly roll pan. A 4-inch offset spatula is useful for smoothing batters or soft doughs in smaller cake or tart pans and for myriad other small tasks. Removing cookies from baking sheets and flipping pancakes is easiest with an ordinary pancake turner/spatula. Spatulas with thin metal blades are much easier to slide under cookies or pancakes than plastic or coated nonstick utensils (which are needed for nonstick cookware and bakeware).

Spatulas—Silicone

Silicone spatulas have replaced rubber spatulas and most wooden spoons in our kitchens. Three sizes—large, medium, and tiny—take care of any contingency. Silicone spatulas can sweep the sides, corners, and bottoms of pots without missing a square inch of territory where contents—whether custard or caramel—may stick or burn. Flat, rather that spoon shaped, spatulas are easier to scrape against the sides of a bowl or pot.

Strainers

A large medium-mesh strainer is good for sifting flour with just one hand and is easier to clean and store than a classic sifter. A medium-fine-mesh strainer is good for straining custards or removing seeds from a berry puree. A fine-mesh strainer (or even a tea strainer) is good for dusting a little powdered sugar or cocoa powder atop a cake or soufflé.

Tart Pans

Tart recipes in this book call for a fluted 9½-inch tart pan with a removable bottom. Tartlet recipes call for fluted 4½-inch pans with removable bottoms. Choose shiny reflective pans rather than those made of dark steel.

Thermometers

Good bakers need reliable ovens. Oven thermometers are inexpensive and are useful for checking the accuracy of your oven temperature.

A standard instant-read meat thermometer with a range from 0°F to 220°F is good when making custards, but recipes that involve cooking caramel or deep-frying require a thermometer with a greater range. Polder makes a thermometer with a range of 32°F to 392°F; it has a remote wire probe that can be attached to a pot or stuck into a chicken in the oven, while the digital display sits on the counter. The excellent Thermapen has a range of 58°F to 592°F and a large display, but no remote probe or clip.

Timer

A timer (whether on your stove, on your phone, or the old-fashioned countdown type with a dial that twists) reminds the busy cook to check the cookies or cake before it's too late.

Whisks

Wire whisks are great for whipping cream and mixing batters as well as blending dry ingredients and fluffing and aerating flour—often eliminating the need to sift.

Zester

A flat (rather than round) Microplane zester shreds and removes the thin colored layer of any citrus peel effortlessly. It is also handy for grating whole nutmeg or cinnamon sticks.

TEN TIPS FOR SUCCESS

Working with new methods and new flours that behave in new or unexpected ways means that the little details are more important than ever! Here are my top ten.

1. Keep flours fresh: Store whole-grain and other perishable flours in airtight containers in the fridge or freezer to prevent them from turning rancid.

2. Do your mise en place like a pro: Measuring all of your ingredients before you start ensures that you have what you need and will not forget an ingredient along the way.

3. Measure accurately: A scale is always the best way to measure ingredients, but if you must use measuring cups, review the proper methods on pages 29 and 42.

4. Check your oven temperature: Put an oven thermometer in the center of the oven and make sure it registers the same temperature that you set. If not, adjust the dial. Better yet, get a professional to calibrate your oven. For convection ovens, consult the manual that came with the oven; you may be instructed to use a lower temperature and check for doneness early.

5. Position your oven racks: Things bake differently on different racks. Unless using a convection oven, always position your oven racks at the level called for in each recipe and rotate pans if called for as well. For convection ovens, consult the oven manual for rack position advice.

6. Mix it right: Some recipes need vigorous beating or whisking; some need only delicate folding. Use the utensil, type of action, and speed called for. Where it matters, mixing times are given to help you get it right. Don't forget to set a timer!

7. Rest and hydrate: Some flours in some recipes need time to absorb liquid from the batter so that the finished cakes or cookies don't taste raw or feel gritty on the palate. Hydration is built into the recipes where necessary. Fruit cobbler recipes may call for mixing the biscuit topping first and letting it rest (and hydrate!) while the fruit is prepared and baked. Cookie recipes may call for resting dough for at least 2 hours, but preferably longer—and I don't call for it unless I know that the payoff is worth the wait.

8. Use the type and size of pans called for: Metal and glass pans bake differently. A 9-inch pan may not seem much bigger than an 8-inch one, but it has 25 percent more surface area, and the same is true for a 9-inch square compared to an 8-inch square. If your pan is too large, your cake will be thinner than you expect and overbaked when your timer goes off.

9. Always cool cakes and cookies completely: Cool them until fully room temperature, not even slightly lukewarm, before frosting or wrapping (unless instructed otherwise).

10. Create like a pastry chef: The best way to make a new successful recipe from an existing but unfamiliar recipe is to first follow the original recipe as written, so that you know what the flavors and textures are like when the recipe works. Then change one thing at a time. If you make too many changes at once, you will never know which change was responsible for the failure.

CUP
1/4 C

CHAPTER ONE

RICE FLOUR

RICE FLOUR is deceptive—seemingly neutral and bland, it can be breathtaking, even exotic, in the simplest of cakes. It is sometimes used for its own subtle, slightly floral flavor, but it more often supports other ingredients. You'll find recipes using three types of rice flour here (and in other chapters in a supporting role); each type has different qualities.

White rice flour is milled from short- or long-grain rice with the bran and germ removed. As such, it is mostly starch and has a very subtle flavor with light floral and cream notes. White rice flour varies in fineness across brands (see page 28 for a discussion)—those from Thailand are so fine, they require separate volume measures. Fineness affects the texture of the finished dessert: finer flour yields a silkier texture in high-moisture dishes such as pudding and ice cream, and also hydrates faster in dishes with less moisture such as scones, giving a less crumbly crumb and better structure. Meanwhile, the ultrafine Thai flour produces very heavy sponge cakes! Rice flour may compact substantially in the bag, which causes volume measures to vary wildly in weight—one more good reason to use a scale for measuring. White rice flour is a perfect supporting actor, amplifying the other flavors in the dish such as butter, cream, milk, eggs, and any bold, assertive flavor like lemon or chocolate. In the chapters that follow, the neutrality of rice flour is used to highlight the flavor of other, more flavorful flours.

Glutinous white rice flour (aka sweet rice flour) is flour milled from sticky, or "sweet," rice and, despite its name, does not contain gluten. Fans of Japanese food may know sweet rice flour as the main ingredient in mochi. It is particularly useful for its ability to retain a satisfying chewiness for days without becoming stale, and it is the secret ingredient that keeps Beignets (page 65) fresh enough to reheat successfully. Recipes in this book call for glutinous white (rather than brown) rice flour.

Brown rice flour has much of the neutrality of white rice flour, plus a light graininess and toasty and delicious caramel flavor. The bran and germ in this whole-grain flour also mitigate white rice flour's tendency toward gumminess. Brown rice flour does perform better with time for hydration—when it is baked immediately in a recipe with a small amount of moisture, it can have a bit of grittiness. Normally typecast as a "health food" ingredient, brown rice flour makes a moist and delicious sponge cake (see page 57) and decadent brownies (see pages 70 and 71).

FLAVOR AFFINITIES FOR RICE FLOURS

Accentuates all other flavors,
including butter, milk, cream, eggs, and chocolate.

WHERE TO BUY AND HOW TO STORE

Rice flours are available in better supermarkets in the specialty flour or gluten-free section of the baking aisle or online from Authentic Foods or Bob's Red Mill (see Resources, page 351). Asian groceries carry Thai rice flour, which is ultra fine and best for a few recipes, but not suitable for every recipe in this book. Brown rice flour is a whole grain and should be stored in an airtight container, away from heat and light, for 2 to 3 months at room temperature, or 6 months in the refrigerator, and up to 12 months in the freezer. White and glutinous white (sweet rice) rice flours are not whole-grain flours, and therefore keep well in a sealed container at room temperature for at least 12 months.

WHITE RICE CHIFFON CAKE

Imagine a cake as light and moist and ethereal as angel food but with a more intriguing flavor and far less sugary sweetness. What a dreamy surprise! Just when I thought rice was simply a neutral background for more interesting or assertive flavors, this cake steals the show with its pristine and subtle rice flavor. Try it with slightly sweetened sliced strawberries and plain or Rose Whipped Cream (page 345), or with sliced mangoes or pineapple. When you want a change, consider adding fragrant cardamom, saffron, or citrus zest. SERVES 10 TO 12

---

1 cup plus 2 tablespoons (224 grams) sugar

5 large egg yolks, at room temperature

¾ cup cool water

½ cup flavorless vegetable oil (such as rice bran, corn, or safflower oil)

1⅓ cups (200 grams) white rice flour
NOTE: This recipe is not successful with Thai white rice flour.

2 teaspoons baking powder

½ teaspoon salt

8 large egg whites, at room temperature

½ teaspoon cream of tartar

EQUIPMENT

Stand mixer with whisk attachment

10-inch tube pan with removable bottom, ungreased

Position a rack in the lower third of the oven and preheat the oven to 325°F.

Set aside ¼ cup (50 grams) of the sugar for later (to stiffen the egg whites).

In a large bowl, whisk the remaining sugar, egg yolks, water, oil, rice flour, baking powder, and salt until thoroughly blended. Set aside.

In the bowl of the stand mixer, beat the egg whites and cream of tartar with the whisk attachment on medium-high speed until the egg whites are creamy white and hold a soft shape when the beaters are lifted. Slowly sprinkle in the reserved sugar, beating on high speed until the egg whites are stiff but not dry. Scrape one-quarter of the egg whites onto the batter and use a rubber spatula to fold them in. Fold in the remaining egg whites. Scrape the batter into the pan and spread it evenly.

Bake for 50 to 55 minutes, until the top of the cake is golden brown and a toothpick inserted in the center comes out clean.

Set the pan on a rack. While the cake is still hot, slide a thin knife or spatula around the sides, pressing against the pan to avoid tearing the cake. Leave the cake in the pan to cool—it will settle at least 1 inch.

When cool, lift the tube to remove the cake. Slide a thin knife or skewer around the tube and slide a spatula under the cake all around. Lift the cake off the bottom of the pan using two spatulas (one on either side of the tube) and transfer it to a serving platter. The cake keeps, wrapped airtight, at room temperature for at least 3 days, or in the freezer for up to 3 months; bring to room temperature before serving. Slice with a serrated knife.

(recipe continues)

VARIATIONS

Lemon Rice Chiffon Cake

Grate the zest of a medium lemon on top of the batter before folding in the egg whites.

Orange Rice Chiffon Cake

Substitute ¾ cup orange juice for the water and grate the zest of 1 medium orange on top of the batter before folding in the egg whites.

Cardamom and Saffron Rice Chiffon Cake

Using a mortar and pestle, pulverize the seeds from 8 cardamom pods (or use ½ teaspoon ground cardamom) and 25 threads of saffron (scant 1⁄16 teaspoon finely chopped or crushed). Add to the batter before folding in the egg whites.

BROWN RICE SPONGE CAKE WITH THREE MILKS

Of course you *could* serve this buttery sponge cake plain (or splashed with a little sweetened espresso or coffee liqueur) and topped with strawberries and whipped cream. But the brown rice flour adds a delicate caramel flavor to the cake, so why not run with it? This riff on the traditional Latin American tres leches—sponge cake drenched in a combo of heavy cream, sweetened condensed milk, and evaporated milk—is less sweet and less drenched than the authentic version, but a terrific variation nonetheless. The warm or cooled cake is poked with a chopstick or the not-too-thick handle of a wooden spoon and then soaked with a sauce of dulce de leche and evaporated milk. The third "milk" is whipped cream on top. More sauce is passed separately at the table. What could be more delicious? SERVES 10 TO 12

FOR THE CAKE

6 tablespoons (85 grams) Clarified Butter (page 335) or ghee (see page 38)

¾ cup (100 grams) brown rice flour, preferably superfine

⅔ cup (130 grams) sugar

4 large eggs

⅛ teaspoon salt

FOR THE SAUCE

1 can (12 ounces) evaporated milk

1 generous cup (350 grams) purchased dulce de leche or cajeta

⅛ teaspoon salt

Whipped Cream (page 343), unsweetened or very lightly sweetened

EQUIPMENT

8-by-2-inch round cake pan

Stand mixer with whisk attachment

Sifter or medium-mesh strainer

Position a rack in the lower third of the oven and preheat the oven to 350°F. Line the bottom of the pan with parchment paper, but do not grease the sides of the pan.

Put the clarified butter in a small pot or microwavable container ready to reheat when needed, and have a 4- to 5-cup bowl ready to pour it into as well—the bowl must be big enough to allow you to fold some batter into the butter later.

Whisk the flour with 2 tablespoons of the sugar in a medium bowl.

Combine the remaining sugar, eggs, and salt in the bowl of the stand mixer and beat with the whisk attachment on high speed for 4 to 5 minutes. The mixture should be light colored and tripled in volume, and you should see well-defined tracks as the whisk spins; when the whisk is lifted, the mixture should fall in a thick, fluffy rope that dissolves slowly on the surface of the batter.

Just before the eggs are ready, heat the clarified butter until very hot and pour it into the reserved bowl.

Remove the bowl from the mixer. Sift one-third of the flour over the eggs. Fold with a large rubber spatula until the flour is almost blended into the batter. Repeat with half of the remaining flour. Fold in the rest of the flour. Scrape about a quarter of the batter into the hot butter. Fold until the butter is completely blended into the batter. Scrape the buttery batter over the remaining batter and fold just until blended. Scrape the batter into the pan.

(recipe continues)

Bake for 30 to 35 minutes, until the cake is golden brown on top. It will have puffed up and then settled level, but it won't have pulled away from the sides of the pan and a toothpick inserted in the center will come out clean and dry. Set the pan on a rack. While the cake is still hot, run a small spatula around the inside of the pan, pressing against the sides of the pan to avoid tearing the cake.

At your convenience (the cake can be warm or completely cool), invert the pan to remove the cake and peel off the parchment liner. Turn the cake right side up. (The cake should be completely cool before storing.) The cake may be wrapped airtight and stored at room temperature for 2 days, or frozen for up to 3 months.

To assemble the cake, an hour (or up to several hours) before serving, set it (cooled or warm) on a rimmed serving platter and poke holes 1 inch apart all over it with a chopstick or a thin wooden spoon handle. Make the sauce by stirring the evaporated milk, dulce de leche, and salt together until smooth. Spoon 1½ to 2 cups of the sauce over the cake, a little at a time, allowing it to be absorbed. Use the greater quantity if you want a more soaked cake, or let some of the delicious cake remain dry. Either way, pass the extra sauce at the table. Make sure the cake is completely cool before topping it with swirls of unsweetened or very lightly sweetened whipped cream, leaving the sides bare. Refrigerate the cake in a covered container or under a cake dome. Leftover cake keeps in the refrigerator for a few days.

LEMON CREAM ROULADE WITH STRAWBERRY-MINT SALAD

This simple cake roll is sweet and tart and yummy to start with—but extra pretty and tempting with its refreshing little salad of minty strawberries and micro-diced jicama (for added crunch). The cake is a very basic white rice sponge (see sidebar, page 63); you could substitute ¾ cup (100 grams) brown rice flour, preferably superfine, for the white rice flour for a less neutral cake with hints of caramel. **SERVES 10 TO 12**

FOR THE SPONGE SHEET

6 tablespoons (85 grams) Clarified Butter (page 335) or ghee (see page 38)

1 cup (150 grams) white rice flour
NOTE: This recipe is not successful with Thai white rice flour.

1 cup (200 grams) granulated sugar

6 large eggs

Scant ¼ teaspoon salt

1 cup Lemon Curd (page 341)

1 cup heavy cream

½ teaspoon pure vanilla extract

Powdered sugar for dusting

FOR THE STRAWBERRY-MINT SALAD

1 pint (230 grams) strawberries

½ cup (65 grams) finely diced (¼-inch) jicama

About ¼ cup (6 grams) loosely packed chopped fresh mint leaves

2 teaspoons sugar, or more to taste

Tiny pinch of salt

Lemon juice

Position a rack in the center of the oven and preheat the oven to 350°F. Line the bottom of the pan with parchment paper, but do not grease the sides of the pan.

Put the clarified butter in a small pot or microwavable container ready to reheat when needed, and have a 4- to 5-cup bowl ready to pour it into as well—the bowl must be big enough to allow you to fold some batter into the butter later.

Whisk the flour and 2 tablespoons of the granulated sugar together thoroughly in a medium bowl.

Combine the remaining granulated sugar, eggs, and salt in the bowl of the stand mixer and beat with the whisk attachment on high speed for 4 to 5 minutes. The mixture should be light colored and tripled in volume, and you should see well-defined tracks as the whisk spins; when the whisk is lifted, the mixture should fall in a thick, fluffy rope that dissolves slowly on the surface of the batter.

Just before the eggs are ready, heat the clarified butter until very hot and pour it into the reserved bowl.

Remove the bowl from the mixer. Sift one-third of the flour over the eggs. Fold with a large rubber spatula until the flour is almost blended into the batter. Repeat with half of the remaining flour. Fold in the rest of the flour. Scrape about a quarter of the batter into the hot butter. Fold until the butter is completely blended into the batter. Scrape the buttery batter over the remaining batter and fold just until blended. Scrape the batter into the pan and spread it evenly with an offset spatula, using as few strokes as possible to avoid deflating the batter.

(recipe continues)

EQUIPMENT

16-by-12-by-1-inch half sheet pan or 11-by-17-inch jelly roll pan

Stand mixer with whisk attachment

Sifter or medium-mesh strainer

Bake for 10 to 15 minutes, or until the top is golden brown and springs back when gently pressed with your fingers. Set the pan on a rack to cool completely before filling.

Run a knife around the edges of the pan to detach the cake. Cover the cake with a sheet of wax paper and set a baking sheet on top. Hold the pans together and flip them over. Remove the top pan and peel the parchment off the sponge. Cover the cake with a sheet of foil, top with the baking sheet, flip the whole business over again, and remove the baking sheet. The cake should be right side up on the sheet of foil.

Spread the surface of the cake evenly with the lemon curd. Whip the cream and vanilla in a chilled bowl until almost stiff. Spread the cream over the lemon curd. Start rolling the cake at one short end by folding the edge of the cake about ½ inch over the cream. Continue to roll the cake, using the foil beneath it to help you. Roll the cake gently but tightly, as though it were a sleeping bag, to keep the roll as cylindrical as possible. When the roll is complete, wrap it in the foil and refrigerate for at least 2 hours or up to 2 days before serving.

Make the salad shortly before serving: Cut the strawberries in half and cut each half into two to four pieces, depending on the size of the berries. In a small bowl, toss the strawberries with the jicama, mint, sugar, salt, and a few drops of lemon juice to taste.

To serve, remove the foil from the roulade and slide the cake onto a platter. Sieve a little powdered sugar over it. Serve slices with a spoonful of the salad.

VARIATION: Individual Lemon Roulades

16-by-12-inch white rice sponge sheet, baked and cooled in the pan

1½ cups Lemon Curd (page 341)

½ cup (55 grams) sliced almonds, toasted

Powdered sugar for dusting

Remove the cake from the pan, peel the liner, and turn the cake right side (brown side) up on a sheet of foil as described on page 62. Set aside ¼ cup of the lemon curd (cover and refrigerate it if not finishing the roulades right away). Spread the rest of the curd over the cake. Cut the cake into 5 equal strips, each about 3¼ inches wide and 11 to 12 inches long. Cut the strips in half so that you have 10 short strips, each 5½ to 6 inches long. Roll each strip into an individual jelly roll. Roulades may be made to this point and refrigerated, covered, for up to 2 days. In any case, they are easier to handle if they have been refrigerated for at least an hour before finishing.

To finish the roulades, spread the ends of each roulade with any lemon curd that may have oozed out, plus some of the reserved curd, as needed. Put the almonds in a flat dish and press the ends of each roulade into them. You can serve the roulades immediately or cover and refrigerate them until you are ready. Just before serving, sieve a little powdered sugar over each roulade.

White Rice Sponge Cake (aka Génoise)

The plain sponge cake known as génoise is the workhorse of classic French baking. It is normally on the dry side, all the better for soaking with flavored syrup and filling with rich buttercreams or mousses. White rice génoise may at first seem like the least interesting cake in this entire collection. But in fact it makes a great all-purpose génoise, because it has a very delicate neutral flavor with a nice hint of butter. In comparison to wheat flour, rice flour allows more butter flavor to come forward so you literally get more butter flavor for your buck. You can substitute it for the traditional plain génoise throughout the classic repertoire of French layer cakes, or gâteaux—well beyond the scope of desserts included in this book. For a slightly softer crumb, try a combination of clarified butter and flavorless vegetable oil. In the preceding roulade recipe, it is used as a thin sponge sheet. To make a round cake, make two-thirds of the recipe and bake for 30 to 35 minutes in an 8-by-2-inch round cake pan lined with parchment on the bottom, sides ungreased. This is a versatile cake indeed.

BEIGNETS

The aroma when frying these beignets is the first clue that they taste gloriously of yeast, butter, and eggs. Powdered or cinnamon sugar is always a good finish, or go overboard and coat them with bittersweet chocolate glaze. Do try the technique for reheating; they are just as good as freshly fried without the last-minute attention. The sweet rice flour holds moisture in these doughnuts and makes them slightly chewy. It also helps them stay fresher longer. **MAKES 3 DOZEN BEIGNETS**

---

2 tablespoons very warm (105°F to 115°F) water

1 tablespoon plus 1 teaspoon granulated sugar

1 teaspoon active dry yeast

½ cup water

4 tablespoons (½ stick/55 grams) unsalted butter

½ teaspoon salt

⅔ cup (100 grams) glutinous rice flour
—OR—
¾ cup plus 2 tablespoons (100 grams) Thai glutinous white rice flour

⅓ cup plus 1 tablespoon (60 grams) white rice flour
—OR—
½ cup plus 1 tablespoon (60 grams) Thai white rice flour

3 large eggs

1 teaspoon pure vanilla extract

1 to 1½ quarts vegetable oil, such as peanut or corn

½ cup (55 grams) powdered sugar

EQUIPMENT

Stand mixer with paddle attachment

Deep-fat fryer or medium (2- to 3-quart) saucepan

Frying thermometer

Combine the warm water, 1 teaspoon granulated sugar, and yeast in a small bowl and set aside.

Combine the 1 tablespoon granulated sugar, ½ cup water, butter, and salt in a small saucepan and bring to a boil over medium heat. Add about half of the glutinous rice flour and all of the white rice flour and stir with a long-handled metal or wooden spoon until smooth. Turn the heat to low and push the dough around the pan for 2 more minutes, turning it over in the pan to avoid scorching. Scrape the dough into the mixer bowl. Break the eggs into the still-hot saucepan and swirl to warm them.

With the mixer on medium speed, add the eggs one at a time, beating after each is added until the dough is glossy and smooth. When all of the eggs are added, scrape down the sides of the bowl and add the yeast mixture, vanilla, and the remaining glutinous rice flour. Mix on medium speed until very smooth and elastic.

Pour oil to a depth of about 2 inches in the deep-fat fryer or saucepan and heat to 350°F. Using two spoons or a small spring-loaded scoop, place 1½-teaspoon-sized lumps of batter in the oil. Do not crowd the pan or fryer; each lump of batter will expand about eightfold. After a minute or so, use long-handled tongs to turn the beignets. Fry until very brown on all sides, 3 to 5 minutes. If necessary to test doneness, cut a beignet in half. Drain on a cake rack; repeat with the remaining batter.

To serve immediately, roll the beignets in powdered sugar (see Note). To serve later, reheat the beignets for 5 minutes in a 400°F oven and then roll in powdered sugar. The beignets may be stored, loosely covered with a paper towel, at room temperature for up to 2 days before reheating.

(recipe continues)

NOTE: The easiest and tidiest way to roll beignets in spiced sugar or powdered sugar without getting your hands in the mix is to pile the beignets in a medium lightweight metal (not plastic) bowl with the sugar and tumble them gently back and forth into another lightweight bowl, until all are coated.

VARIATIONS

Beignets with Cinnamon Sugar

Substitute 2 teaspoons ground cinnamon mixed with ½ cup (100 grams) granulated sugar for the powdered sugar.

Chocolate-Glazed Beignets

Omit the powdered sugar and dip warm beignets in chocolate glaze: Boil 1¼ cups water and 2¼ cups (450 grams) sugar together until the sugar dissolves. Add 3 cups (510 grams) semisweet or bittersweet chocolate chips all at once; let rest for 5 minutes, then add 3 tablespoons vegetable oil and stir until smooth. Use a slotted spoon to dip the beignets in the glaze and set them on a rack until the glaze is set.

ALMOND TUILES

Crispy, crunchy almond cookies are elegantly thin and buttery. They partner perfectly with creamy desserts like custards, pudding, and ice cream. Rice flour plays a neutral role here, allowing the flavor of butter and almonds to shine. Classic tuiles are cooled over a rolling pin (see the box on page 69 for methods, including a shortcut) to make them resemble the roof tiles they are named for, but you can also skip that step and make them flat. **MAKES ABOUT FORTY 3-INCH COOKIES**

---

4 tablespoons (½ stick/55 grams) unsalted butter, melted, plus more for greasing the foil

2 large egg whites

2 teaspoons water

½ cup (100 grams) sugar

3 tablespoons (30 grams) white rice flour

—OR—

Scant ⅓ cup (30 grams) Thai white rice flour

¼ teaspoon pure almond extract

Scant ½ teaspoon salt

⅔ cup (70 grams) sliced almonds

EQUIPMENT

2 baking sheets

Silicone baking mats (optional)

Rolling pin or small cups for shaping (optional); see page 69

Line the baking sheets with regular foil (dull side facing up) or silicone mats and grease the foil or silicone lightly but thoroughly, or line the sheets with nonstick foil, nonstick side up (see Note).

In a medium bowl, mix the egg whites with the water, sugar, rice flour, almond extract, and salt until well blended. Stir in the butter and almonds. Cover the bowl and let the batter rest for several hours or overnight in the refrigerator to let the flour absorb moisture.

Position racks in the upper and lower thirds of the oven (or one rack in the center if you are baking only one sheet at a time) and preheat the oven to 325°F.

Stir the batter. Drop level teaspoons 2 inches apart on the lined baking sheets. Use the back of the spoon to smear the batter into 2½-inch rounds. Bake, watching carefully, for 12 to 15 minutes, rotating the sheets from front to back and top to bottom about halfway through the baking time, until the tuiles are mostly deep golden brown. If the cookies are not baked enough, they will not be completely crisp when cool, nor will they come off the foil easily.

Slide the foil sheets onto racks and let the cookies cool completely before removing them. Or, if using silicone mats, transfer the hot cookies to a rack to cool.

To retain crispness, put the cookies in an airtight container as soon as they are cool. They may be stored airtight for at least 1 month.

(recipe continues)

NOTE: Tuile batter baked on greased foil spreads in the oven, making slightly irregular-shaped cookies with thin, delicate edges. The batter will not spread on nonstick foil or silicone mats, so you must spread it to the diameter that you want before the sheets go into the oven. The edges of the tuiles will not be as delicate as those baked on foil, but they will hold their shapes.

Shaping Tuiles

To make traditional curved tuiles, you must shape the cookies while they are still hot, so you might want to bake only one sheet at a time until you get the hang of it. For silicone mats: Slide a small metal spatula under each cookie immediately, or as soon as you can do so without deforming them. Drape cookies over a rolling pin (anchored so it will not roll) or into custard cups or any small container that will give them an interesting shape. Move the cookies to a cooling rack when they are cool enough to keep their shape. Repeat until all of the tuiles are shaped. (If the cookies become too brittle, return them to the oven for a couple of minutes until they are hot and flexible again.)

Tuiles baked on regular foil tend to stick until they are completely cool, so use this trick for shaping a whole sheet at a time: Grasp the edges of the foil as soon as the sheet comes from the oven (without touching the hot pan or the cookies) and roll the foil into a fat cylinder, gently curving the attached cookies like potato chips. Crimp or secure the foil with a paper clip. When cool, unroll the foil carefully and remove the tuiles.

ULTRA-BITTERSWEET BROWNIES

My own "notes to self" declared these decadent bars "spectacular while still warm" and "superb," with "crunchy crust at the edges and a little crunchy/chewy on top but creamy and gooey inside." Can it get any better than that? It must be emphasized nonetheless that these are intensely chocolate brownies and not necessarily for kids (see the next recipe for a kid-friendly brownie). I cut them into twenty-five pieces instead of sixteen. You will need to use a mixer—the handheld type is perfect here—to be sure that the ultrarich batter comes together smoothly. **MAKES 25 SMALL BROWNIES**

---

1 pound 55% to 60% chocolate (see Note for other chocolate choices), coarsely chopped

8 tablespoons (1 stick/115 grams) unsalted butter, cut into chunks

1 cup (200 grams) sugar

2 teaspoons pure vanilla extract

Scant ½ teaspoon salt

3 large eggs, cold

½ cup plus 1 tablespoon (70 grams) brown rice flour

EQUIPMENT

Handheld mixer

8-inch square baking pan, lined on the bottom and all four sides with foil

NOTE: You can use 13 ounces of 62% to 64% chocolate instead.

Position a rack in the lower third of the oven and preheat the oven to 350°F.

Place the chocolate and butter in a medium stainless steel bowl set directly in a wide skillet of barely simmering water. Stir frequently until the chocolate is melted and the mixture is smooth and hot to the touch. Remove the bowl from the skillet and stir in the sugar, vanilla, and salt. Let the mixture cool to lukewarm. Add the eggs one at a time, beating with the handheld mixer after each addition until incorporated. Add the flour and beat on medium speed until the batter is smooth (it should not look curdled or separated), slightly lightened in color, and beginning to come away from the sides of the bowl, 1 or 2 minutes or more.

Scrape the batter into the pan and spread it evenly but with lots of raised swirls and ridges—these look great and get slightly crusty in the oven. Bake for 30 to 35 minutes, until the surface looks glossy but dry, with a few hairline cracks, and a toothpick inserted in the center comes out with a few gooey crumbs on it.

Cool in the pan on a rack. Lift the foil edges to transfer the brownies to a cutting board. Cut into 25 squares. The brownies keep in an airtight container for 2 to 3 days.

ALMOND AND BROWN RICE BROWNIES

Much more kid friendly than Ultra-Bittersweet Brownies (opposite), these brownies made with a combination of brown rice flour and almond flour are quite a bit less intense, but still dark, gooey, and super chocolaty. **MAKES SIXTEEN 2-INCH BROWNIES**

½ cup (70 grams) whole almonds, or ¾ cup (70 grams) almond flour/meal

¼ cup plus 2 tablespoons (50 grams) brown rice flour

6 ounces (170 grams) 60% to 62% chocolate, coarsely chopped

6 tablespoons (85 grams) unsalted butter, cut into chunks

½ teaspoon salt

⅔ cup (130 grams) sugar

1 teaspoon pure vanilla extract

2 large eggs, cold

1 cup (100 grams) walnut or pecan pieces (optional)

EQUIPMENT

Food processor fitted with the steel blade (optional)

8-inch square metal baking pan, lined on the bottom and all four sides with foil

Position a rack in the lower third of the oven and preheat the oven to 325°F.

If using whole almonds, put them in the food processor with the rice flour and pulse until the nuts are finely ground. If using almond flour, simply mix it in a bowl with the rice flour. Set aside.

Melt the chocolate with the butter in a medium stainless steel bowl set directly in a wide skillet of barely simmering water. Stir frequently until the mixture is melted and hot to the touch.

Remove the bowl and stir in the salt, sugar, and vanilla. Let cool until the mixture is lukewarm. Stir in the eggs one at a time. Add the almond-flour mixture and stir until moistened, then mix briskly for about 40 strokes. Stir in the walnuts or pecans, if using.

Scrape the batter into the prepared pan and spread it evenly. Bake for 20 to 25 minutes, or until the brownies are slightly puffed all over and a toothpick inserted in the center comes out moist but clean.

Cool in the pan on a rack. Lift the foil edges to transfer the brownies to a cutting board. Cut into 16 squares. The brownies keep in an airtight container at room temperature for 2 or 3 days.

BRANDIED CHERRY CLAFOUTIS

Clafoutis always seemed like bland custard slightly curdled by watery fruit, or soggy cake with equally soggy cherries. But Maya Klein's version is a game changer: a silken filling laced with intensely flavorful fruit and just a bit of toothsome crustiness at the edge. **SERVES 6 TO 8**

¼ cup brandy or 1 tablespoon lemon juice

¼ cup water

¾ cup (150 grams) sugar

1⅓ cups (225 grams/8 ounces) dried cherries (sour or Bing)

⅓ cup plus 1 tablespoon (60 grams) white rice flour
—OR—
½ cup plus 1 tablespoon (60 grams) Thai white rice flour

Pinch of salt

2 large egg yolks

1 large egg

1 cup milk

1 teaspoon pure vanilla extract

3 tablespoons (45 grams) unsalted butter

EQUIPMENT

9-inch glass or ceramic pie dish or other 1-quart baking dish

NOTE: Letting the fruit linger briefly in the hot pan without batter concentrates flavor and eliminates excess moisture. Served warm, the clafoutis retains a bit of puff from baking, but chilling it for a day or so accentuates the buttery quality of the custard

In a small saucepan, combine the brandy, water, and ¼ cup of the sugar. Warm over medium heat. After 1 or 2 minutes, ignite the alcohol: On a gas range, stand back, turn the heat to high, and bring the edge of the pan near the flame. On an electric range, use a long match. Allow the flames to burn down and add the cherries. (If using lemon juice, warm to a simmer but do not ignite.) Tumble the cherries in the liquid, cover, and remove from the heat.

Position a rack in the lower third of the oven. Set a 9-inch glass pie dish on the rack and preheat the oven to 350°F.

Meanwhile, whisk the flour, remaining ½ cup sugar, salt, egg yolks, and whole egg in a medium mixing bowl until smooth. Add the milk and vanilla and whisk until well blended. Drain the cherries and discard the liquid. When the oven is preheated, put the butter in the pie dish and swirl to coat the bottom. Scatter the cherries in the dish and return it to the oven for 2 minutes (see Note). Whisk the batter, remove the dish from the oven, and pour the batter into it in a spiral from the outside edge to the center. Bake for 20 to 25 minutes, until lightly browned at the edges and a little bubbly at the center. Remove from the oven and let cool for at least 20 minutes or up to 2 hours before serving. Leftovers can be covered and refrigerated for up to 3 days; warm in the microwave before serving.

VARIATION

Fresh Cherry or Pear Clafoutis

Cherries are classic, but pears have a lovely light, almost floral flavor. Use the lemon juice instead of brandy and omit the water. Do not heat or try to flambé the lemon juice—just toss it with the fruit and the ¼ cup sugar. Substitute 1½ cups (225 grams/ 8 ounces) fresh pitted cherries or cored, cubed ripe pears for the dried cherries.

DARK CHOCOLATE SOUFFLÉS

It can't be said too many times: soufflés are easier to make than anyone thinks, *and* you can prepare them in advance—except for the baking—and simply pop them into a hot oven a few minutes before you are ready to serve. Always call guests to the table early enough to wait for the soufflés to arrive directly from the oven. Guests—unlike soufflés—can wait without deflating!

Rice flour makes the base for a superior chocolate soufflé. The delicate rice flavor allows all of the flavors of the chocolate to come through bright and clear; wheat flour is more traditional but it doesn't work as well because it blocks and mutes other flavors. For years I've made my chocolate soufflés without any flour at all, compromising texture slightly in pursuit of more and better chocolate flavor. Rice flour offers the best of both worlds: it gives the creamy, luxurious texture that comes from using a little flour in the soufflé base, but it doesn't compete with the flavor of an excellent chocolate. **SERVES 7 OR 8**

Unsalted butter, softened, to butter the soufflé cups

Granulated sugar, to coat the soufflé cups

1 tablespoon (15 grams) unsalted butter

½ cup plus 1 tablespoon milk

1 tablespoon (10 grams) white rice flour

⅛ teaspoon salt

8 ounces (225 grams) 60% to 62% chocolate, chopped medium fine

3 large egg yolks, at room temperature

1 teaspoon pure vanilla extract

4 large egg whites, at room temperature

¼ teaspoon cream of tartar

⅓ cup (65 grams) granulated sugar

2 to 3 tablespoons powdered sugar for dusting (optional)

Lightly sweetened whipped cream (see page 343)

Butter the bottom, sides, and rim of the soufflé cups lightly but thoroughly. To coat with sugar, fill one of them with the sugar. Tilt the cup and rotate it over a second cup until the bottom, sides, and rim of both are completely coated with sugar. Pour excess sugar into the third cup and repeat until all cups are coated. Discard any excess sugar (or use it in your coffee or cereal), but don't use it in the egg whites.

In a small saucepan, melt the 1 tablespoon butter and add the ½ cup milk, rice flour, and salt. Whisking constantly, bring the mixture to a simmer and simmer gently for about 1½ minutes to cook the flour. Remove from the heat and add the chocolate. Stir until the chocolate is completely melted and the mixture is smooth. Scrape the mixture into a large bowl and whisk in the egg yolks, vanilla, and remaining 1 tablespoon milk. Set aside.

Combine the egg whites and cream of tartar in the bowl of the stand mixer (or in another large bowl if using a handheld mixer). Beat with the whisk attachment on medium speed (or on high speed with the handheld mixer) until the egg whites are creamy white and hold a soft shape when the beaters are lifted. Gradually sprinkle in the ⅓ cup sugar and continue to beat on high speed until the egg whites are stiff but not dry. Fold one-quarter of the egg whites into the chocolate mixture to lighten it, then fold in

EQUIPMENT

Seven or eight 5- to 6-ounce soufflé cups

Stand mixer with whisk attachment, or handheld mixer

Rimmed baking sheet

the remaining egg whites. Divide the mixture evenly among the sugared cups, filling them at least three-quarters full. The soufflés may be prepared to this point, covered, and refrigerated for up to 3 days before serving.

Position a rack in the lower third of the oven and preheat the oven to 375°F. Set the cups on the baking sheet and bake the soufflés for 15 to 17 minutes, until a bamboo skewer plunged into the center tests moist but not completely gooey or runny. The soufflés will rise and may crack on top before they are done.

Remove the soufflés from the oven and lightly sift powdered sugar over them, if desired. Serve immediately with the whipped cream.

SILKY BUTTERSCOTCH PUDDING

This old-time classic dessert is especially yummy made with rice flour instead of the traditional cornstarch (see box). The scotch is a great flavor addition, too, but you can omit it if you like. Use superfine or Thai rice flour here or the pudding will not be silky smooth. If you can wait, the pudding is even more delicious and silky on the second day. **SERVES 8**

1⅓ cups (270 grams) packed brown sugar

¼ cup scotch or bourbon whiskey *or* 3 tablespoons of water

¼ cup (40 grams) superfine white rice flour

—OR—

⅓ cup plus 1 tablespoon (40 grams) Thai rice flour

¼ teaspoon salt

4 cups half-and-half (or 3½ cups whole milk and ½ cup heavy cream)

EQUIPMENT

Eight 6-ounce custard cups or ramekins

Combine the brown sugar and scotch (or water) in a medium saucepan. Cook over medium heat until the mixture bubbles all over and the sugar is dissolved. Remove the pan from the heat and allow to cool for 20 minutes. Add the rice flour, salt, and about ¼ cup of the half-and-half. Whisk until there are no lumps of rice flour and then whisk in the remaining half-and-half. Using a silicone spatula or a wooden spoon, stir the mixture constantly over medium heat, scraping the bottom, sides, and corners of the pan until the pudding thickens and begins to bubble. It may look curdled at this point; that is okay. Adjust the heat as necessary to maintain a gentle boil. Continue cooking and stirring for 4 more minutes to fully cook the rice flour.

Take the pan off the heat and immediately pour the pudding into the cups. Let the pudding cool at room temperature for 1 hour, undisturbed (without mixing, jiggling, or spooning out a taste). Cover and refrigerate for at least several hours or (better yet) for 24 hours before serving.

Using Rice Flour as a Thickener

Cornstarch is the common thickener for puddings. But in the case of Silky Butterscotch Pudding, Silky Saffron Rice Pudding (page 78), Silky Chocolate Pudding (page 80), Silky Vanilla Pudding (page 78), Sicilian Chocolate Gelato (page 81), and The New Vanilla Pastry Cream (page 342), we use rice flour to do the same job. Superfine rice flour requires a little more simmering time to get rid of its raw starch flavor than does cornstarch, but it gives us a brighter, cleaner flavor of butterscotch—or chocolate, saffron, or vanilla!

SILKY SAFFRON RICE PUDDING

Saffron makes an otherwise cozy pudding exotic and beautiful to look at, especially when garnished with Cardamom Brittle (page 338) or with chopped pistachios and a little grated cinnamon stick. Vanilla works for everyone: serve it to the family with a favorite peanut butter or chocolate cookie or just top it with a spoonful of fruit preserves; dress it up for company with Oat and Almond Tuiles (page 119), Almond Tuiles (page 67), or Crispy Coconut Wafers (page 324). For the most luxurious texture, make these puddings a day ahead. This is one of the few recipes that requires superfine or Thai rice flour; get it from either an Asian grocery store or Authentic Foods (see Resources, page 351). **SERVES 8**

---

30 saffron threads, or 1⁄16 teaspoon finely chopped or crushed

⅔ cup (130 grams) sugar

¼ cup (40 grams) superfine white rice flour
—OR—
⅓ cup plus 1 tablespoon (40 grams) Thai white rice flour

Scant ¼ teaspoon salt

4 cups half-and-half (or 3½ cups whole milk and ½ cup heavy cream)

EQUIPMENT

Eight 6-ounce custard cups or ramekins

Use your fingers to pinch the saffron with some of the sugar to make the particles smaller, or use a mortar and pestle. Whisk the sugar, saffron, rice flour, and salt in a heavy medium saucepan. Add about ¼ cup of the half-and-half and whisk to form a smooth paste. Whisk in the remaining half-and-half. Using a silicone spatula or a wooden spoon, stir the mixture constantly over medium heat until it is very hot to the touch. Remove from the heat, cover, and let steep for 20 minutes. Reheat the mixture, scraping the bottom, sides, and corners of the pan, until the pudding thickens and begins to bubble. Adjust the heat as necessary to maintain a gentle boil. Continue cooking and stirring for 4 more minutes to fully cook the rice flour.

Take the pan off the heat and immediately pour the pudding into the cups. Let the pudding cool at room temperature for 1 hour, undisturbed (without mixing, jiggling, or spooning out a taste). Cover and refrigerate for at least several hours or (better yet) for 24 hours before serving.

VARIATION: Silky Vanilla Pudding

Omit the saffron and the steeping step. Stir 1 teaspoon pure vanilla extract into the pudding as soon as it's cooked. Or use a vanilla bean: Add the seeds scraped from 1 split vanilla bean, and the pod, to the pot with the cream. Heat and steep as directed. Remove the pod (rinse and save for another use or discard) and proceed as directed.

SILKY CHOCOLATE PUDDING

Rice flour, instead of cornstarch, lets the flavors of chocolate and cocoa really sing in this simple but stunning chocolate pudding (see box, page 77). For best flavor, the cocoa and chocolate are added to the pudding only in the last minute or so of cooking. This is one of the few recipes that requires superfine or Thai rice flour, either from an Asian grocery store or from Authentic Foods (see Resources, page 351). SERVES 8

½ cup less 1 tablespoon (40 grams) unsweetened cocoa powder

4 cups whole milk

6 ounces 55% to 64% dark chocolate, finely chopped

⅔ cup (130 grams) sugar

2 tablespoons (20 grams) superfine white rice flour
—OR—
3 tablespoons (20 grams) Thai white rice flour

¼ teaspoon salt

½ cup heavy cream

2 teaspoons pure vanilla extract

1 cup heavy cream, or lightly sweetened whipped cream (see page 343), for pouring (optional)

EQUIPMENT

Eight 4-ounce custard cups or ramekins

In a small bowl, whisk the cocoa and about ¼ cup of the milk to form a smooth, loose paste. Dump the chopped chocolate on top and set aside near the stove.

Whisk the sugar, rice flour, and salt in a heavy medium saucepan. Whisk in a few tablespoons of the remaining milk to form a smooth paste. Whisk in the remaining milk and the ½ cup cream. Using a silicone spatula or a wooden spoon, stir the mixture constantly over medium heat, scraping the bottom, sides, and corners of the pan, until the mixture thickens and begins to bubble at the edges. Set a timer for 2 minutes and continue to cook and stir, adjusting the heat so that the mixture bubbles readily but not furiously.

Scrape about one-third of the mixture over the cocoa and chocolate and whisk until the chocolate is melted. Stir everything back into the saucepan and return it to the heat, whisking constantly. When the pudding starts to bubble at the edges, continue to cook and stir for 1 minute longer. Remove from the heat and stir in the vanilla. Divide the pudding among the cups or ramekins. Serve warm or at room temperature or chilled, with poured cream or whipped cream, if desired. The pudding can be covered and refrigerated for up to 3 days.

VARIATION: Silky Milk Chocolate Pudding

Reduce the amount of cocoa powder to 2 tablespoons (12 grams) and substitute 7 ounces (200 grams) finely chopped milk chocolate for the dark chocolate. Proceed as directed but scrape about half rather than a third of the hot pudding mixture over the chopped chocolate.

SICILIAN CHOCOLATE GELATO

Sicilian gelato is perfect for sultry weather when you crave ice cream but really don't want the fat or calories that go along with it. With no eggs or cream, Sicilian gelato is essentially churned frozen pudding—made with milk and usually thickened with cornstarch. Since rice flour works so well in Silky Chocolate Pudding (opposite)—and all of the other puddings in this chapter—I knew the same substitution would work fabulously for gelato. This recipe has a two-for-one bonus: you can freeze and serve the gelato base without putting it in the ice cream machine for a superrich dark smooth gelato, or you can chill the base as usual and freeze it in your ice cream machine for a lighter texture and flavor. Either way, this is one of the few recipes that requires superfine or Thai rice flour, from either an Asian grocery store or Authentic Foods (see Resources, page 351). **MAKES ABOUT 1 QUART/SERVES 6 TO 8**

½ cup plus 2 tablespoons (60 grams) unsweetened natural or Dutch-process cocoa powder

3 cups whole milk

1 cup (200 grams) sugar

1½ tablespoons (15 grams) superfine white rice flour
—OR—
2½ tablespoons (15 grams) Thai white rice flour

⅛ teaspoon salt

3½ ounces (100 grams) 66% to 72% dark chocolate, finely chopped

EQUIPMENT

Ice cream maker

In a medium bowl, whisk the cocoa with about ⅓ cup of the milk to form a smooth, loose paste. Set aside near the stove.

In a medium saucepan, mix the sugar with the rice flour and salt. Whisk in a few tablespoons of the remaining milk to form a smooth paste. Whisk in the rest of the milk. Cook over medium heat, stirring constantly with a silicone spatula or a wooden spoon, scraping the bottom, sides, and corners of the pan to avoid scorching, until the mixture thickens and bubbles a little at the edges. Continue to cook, stirring constantly, for 3 minutes longer. Scrape the cocoa paste into the pot and whisk to blend. Continue to stir just until a few bubbles appear around the edge of the pot, another minute or so. Add the chocolate and stir until smooth. Let cool.

Cover the surface of the mixture with plastic wrap and chill for several hours or overnight. Freeze without churning for a rich, dense gelato, or freeze with your ice cream maker according to the manufacturer's instructions.

CHAPTER TWO

OAT FLOUR

OATS

OATS flourish in temperate regions such as northwest Europe and North America—and now in the foothills of the Himalayas—as they require less summer heat and tolerate more rain than most cereals.

Since ancient times, oats have been used as much for animal feed as for human food. Samuel Johnson noted in his diary that oats were "eaten by the people of Scotland, but fit only for horses in England." We humans still consume oats mostly in our morning porridge, though we love our oatmeal cookies, muffins, scones, crackers, and hearty breads.

Today, oats are celebrated for their health-giving attributes: Oats have more fat and protein but fewer carbs and less sugar than whole wheat, and they are a good source of calcium. Oats lower total cholesterol and contain unique antioxidants to reduce the risk of cardiovascular disease, stabilize blood sugar, lower the risk of type 2 diabetes, and protect against breast cancer. That morning oatmeal is good for us!

Oat flour is nothing more than oats milled to a fine, soft tan powder; it has a sweet, slightly toasty aroma and a tendency to clump. Oat flour has all the flavor of oats without the texture of oatmeal. Its soft fiber absorbs moisture well, so it doesn't produce the gritty or sandpapery textures that result from some other whole-grain flours.

Chocolate chip cookies made with oat flour were my first clue that oats might actually taste like butterscotch; initially I thought that the oats accentuated the flavor of the brown sugar in the cookies. It took making a sponge cake with oat flour (and not a bit of brown sugar) to realize that the oats themselves had contributed butterscotch flavor!

Oat flour makes tender, light cakes and toothsome, even delicate, cookies. Many of these recipes use oat flour alone, with stellar results, in cakes, crumbles, ginger cookies, and crispy wafers called tuiles. Teamed up with white or brown rice flour in other recipes, oat flour softens the texture, reduces the grit, or just adds a little complexity to the flavor. The Ultimate Butter Cake (page 90), Butter Biscuits (page 109), and Simple Scones (page 107), otherwise rice-based, taste more interesting with a little oat flour, though one would be hard put to identify the oat flavor if not looking for it. Oat flour with rice flour also makes a sensational American-style chocolate fudge cake (see page 103)—but then, why wouldn't an undertone of butterscotch enhance the flavor of chocolate?

Don't be surprised if guests don't immediately recognize the flavor of oats in most of these recipes, or if friends who think they don't like oats love these desserts. Oats themselves are gluten free, but may be contaminated from adjacent wheat fields. Look for oats and oat flour in packages labeled "gluten free" to avoid that possibility.

FLAVOR AFFINITIES FOR OAT FLOUR

Nuts, brown sugar, caramel, honey, maple sugar, butter, fresh apples, blueberries, bananas, pears, figs, dates, raisins, prunes, cinnamon, nutmeg, yogurt, cream, mascarpone, coconut, coffee, vanilla

WHERE TO BUY AND HOW TO STORE

Oat flour is available in better supermarkets in the specialty flour or gluten-free section of the baking aisle, or online from Bob's Red Mill (see Resources, page 351). Oat flour is a whole grain and should be stored in an airtight container, away from heat and light, for 2 to 3 months at room temperature, or 6 months in the refrigerator, and up to 12 months in the freezer.

OAT FLOUR SPONGE CAKE

Oat flour turns a plain-Jane sponge cake into something elegant with the subtle but distinct flavors of butterscotch or toffee. The crusty edges I picked off my first oat flour sponge cake were delicious with my coffee: you'll find it a perfect match to serve with Prunes Poached in Coffee and Brandy (page 336). Or use this base to make the two-layer Oat Flour Fruit Basket Cake that follows (page 88). **SERVES 8 TO 10**

---

3 tablespoons (45 grams) Clarified Butter (page 335) or ghee (see page 38)

1 cup (100 grams) gluten-free oat flour

⅔ cup (130 grams) sugar

4 large eggs

⅛ teaspoon salt

EQUIPMENT

8-by-3-inch springform pan or cheesecake pan with removable bottom

Stand mixer with whisk attachment

Sifter or medium-mesh strainer

Position a rack in the lower third of the oven and preheat the oven to 350°F. Line the bottom of the pan with parchment paper but leave the sides ungreased.

Put the clarified butter in a small pot or microwavable container ready to reheat when needed, and have a 4- to 5-cup bowl ready to pour it into as well—the bowl must be big enough to allow you to fold some batter into the butter later.

Whisk the flour and 2 tablespoons of the sugar together thoroughly in a medium bowl.

Combine the remaining sugar, eggs, and salt in the bowl of the stand mixer and beat with the whisk attachment on high speed for 4 to 5 minutes. The mixture should be light colored and tripled in volume, and you should see well-defined tracks as the whisk spins; when the whisk is lifted, the mixture should fall in a thick, fluffy rope that dissolves slowly on the surface of the batter.

Just before the eggs are ready, heat the clarified butter until very hot and pour it into the reserved bowl.

Remove the bowl from the mixer. Sift one-third of the flour over the eggs. Fold with a large rubber spatula until the flour is almost blended into the batter. Repeat with half of the remaining flour. Fold in the rest of the flour. Scrape about a quarter of the batter into the hot butter. Fold until the butter is completely blended into the batter. Scrape the buttery batter over the remaining batter and fold just until blended. Scrape the batter into the pan.

Bake until the cake is golden brown on top, 30 to 35 minutes. It will have puffed up and then settled level, but it won't have pulled away from the sides of the pan, and a toothpick inserted in the center should come out clean and dry. Set the pan on a rack.

While the cake is hot, run a spatula around the inside of the pan, pressing against the sides of the pan to avoid tearing the cake.

At your convenience (the cake can be warm or completely cool), invert the pan to remove the cake and peel off the parchment liner. Turn the cake right side up to finish cooling. It should be completely cool before filling, frosting, or storing. The cake may be wrapped airtight and stored at room temperature for 2 days, or frozen for up to 3 months.

OAT FLOUR FRUIT BASKET CAKE

Too simple for words: soft, tender layers of oat flour génoise are filled with preserves, whipped cream, and fresh berries. Team strawberry preserves with fresh strawberries, or pair them with apricot or peach preserves instead. In winter, swap the berries for diced bananas. It will be hard not to eat leftovers for breakfast—but there's nothing wrong with oats and fruit for breakfast. **SERVES 10 TO 12**

---

1 cup heavy cream

1 teaspoon pure vanilla extract

2 to 3 teaspoons granulated sugar

Oat Flour Sponge Cake (page 86), baked and cooled

¼ to ⅓ cup fruit preserves

1½ pints (425 grams) blackberries or raspberries, or 1½ pints (340 grams) strawberries

Powdered sugar for dusting

Lightly sweetened whipped cream (see page 343; optional)

EQUIPMENT

Electric mixer or whisk

Whip the cream with the vanilla in a chilled bowl until it begins to thicken. Add granulated sugar to taste, beating until the cream holds a good shape without being quite stiff—it will continue to stiffen as you spread it on the cake. Refrigerate the cream while you prepare the cake.

Turn the cake best-looking side up on a platter. Cut it into two layers with a serrated bread knife. If the top layer is too delicate to pick up without breaking, slide a rimless baking sheet or a flexible plastic cutting mat under it and set it aside. Spread the bottom layer evenly with the preserves. Spread all of the whipped cream over the preserves. Set aside a few berries for garnish. Cut strawberries into bite-size pieces. Arrange berries or berry pieces over the cream in a single layer with a little space between them. Press the berries well into the cream (so the top cake layer will make contact with the cream). Set the top cake layer on top of the cream and press gently to level the cake. Cover and refrigerate for at least 2 hours or up to 1 day.

Sieve a little powdered sugar over the top of the cake and garnish with the reserved berries and whipped cream, if desired, before serving. Leftovers keep in an airtight container in the refrigerator for a day or so.

ULTIMATE BUTTER CAKE

Lighter than a pound cake but more substantial than a sponge or chiffon cake, this cake *will* become your new basic. The combination of rice flour and a little oat flour for depth of flavor and complexity makes the cake taste extra buttery. The recipe makes a single tube (or Bundt) cake or two cake layers with a perfect melt-in-your-mouth crumb. The cake is excellent without any frosting at all, or with just about any frosting or glaze you can think of. Make it once and you will think of a million ways to use it, flavor it, or style it, from cupcakes to layer cakes—well beyond the variations that follow. **SERVES 10 TO 12**

2⅔ cups (400 grams) white rice flour
—OR—
4 cups (400 grams) Thai white rice flour

½ cup (50 grams) gluten-free oat flour

2 cups minus 3 tablespoons (360 grams) sugar

½ pound (2 sticks/225 grams) unsalted butter, very soft

¾ teaspoon salt

2 teaspoons baking powder

1 teaspoon baking soda

½ teaspoon xanthan gum

1 cup plain yogurt (any percent fat) or slightly watered down Greek yogurt

4 large eggs

2 teaspoons pure vanilla extract

EQUIPMENT

Two 9-by-2-inch round cake pans or one 10-inch tube pan with removable bottom or 10- to 12-cup Bundt pan

Stand mixer with paddle attachment

Position a rack in the lower third of the oven and preheat the oven to 350°F. Grease the pan(s) with vegetable oil spray or butter and line the bottoms of the layer cake pans with parchment paper.

Combine the rice and oat flours, sugar, butter, and salt in the bowl of the stand mixer and mix on medium speed with the paddle attachment until the mixture is the texture of brown sugar, about a minute. Add the baking powder, baking soda, xanthan gum, yogurt, eggs, and vanilla and beat on medium-high speed for 2 to 3 minutes; the batter should be very smooth and fluffy. Scrape into the prepared pan(s) and bake the layers for 25 to 30 minutes, or the tube or Bundt pan for 45 to 50 minutes, until a bamboo skewer inserted in the center comes out clean and dry. Set the pan(s) on a rack to cool completely.

Slide a thin knife or a small metal spatula around the edges of the layer cakes or the tube pan (and the tube) to detach the cake(s) from the pan(s). Loosen the cake from the Bundt pan by rapping all sides of the pan against the counter. Invert each cake onto the rack and peel off the parchment liner. Turn the layer cakes right side up.

VARIATIONS

Orange Butter Cake

Substitute 1 tablespoon plus 1 teaspoon finely grated orange zest and ¼ cup orange juice for the vanilla extract and bake in two layer pans. Spread the cooled layers with apricot jam and serve with lightly sweetened whipped cream (see page 343).

Holiday Pound Cake with Bourbon Glaze

Add 2 teaspoons grated nutmeg to the batter and bake in the tube or Bundt pan. When the cake is cool, mix 1 cup (115 grams) powdered sugar with ¼ cup bourbon (or brandy or rum). Pour the glaze over the cake.

Cupcakes

Line a cupcake tin with 12 paper liners and put 2 extra liners in 2 custard cups or ramekins. Make half of the recipe for the butter cake or the orange cake and fill the cups about two-thirds full; there will be exactly enough batter for 14 cupcakes and they will overflow if you put all of the batter in 12. Bake for 20 to 25 minutes, until a toothpick inserted near the center comes out almost clean. Let cool and frost as desired.

POPPY SEED POUND CAKE

Preparing the pan with poppy seeds gives a nutty-tasting, crunchy crust and guaranteed pan release. The cake is moist and just gets better over a few days. If it gets a little stale, toast slices and serve them with some English marmalade. **SERVES 8 TO 12**

½ pound (2 sticks/225 grams) unsalted butter, very soft

¼ cup plus 2 tablespoons (60 grams) poppy seeds

2⅔ cups (400 grams) white rice flour
—OR—
4 cups (400 grams) Thai white rice flour

½ cup (50 grams) gluten-free oat flour

2 cups minus 3 tablespoons (360 grams) sugar

¾ teaspoon salt

2 teaspoons baking powder

1 teaspoon baking soda

½ teaspoon xanthan gum

1 cup plain yogurt (any percent fat) or slightly watered down Greek yogurt

4 large eggs

2 teaspoons pure vanilla extract

½ teaspoon pure almond extract

EQUIPMENT

10- to 12-cup Bundt pan

Stand mixer with paddle attachment

Position a rack in the lower third of the oven and preheat the oven to 350°F. Use about 2 tablespoons of the butter to heavily coat the inside of the Bundt pan, making sure not to leave any bare spots. Sprinkle about 2 tablespoons of the poppy seeds into the bottom of the pan, then shake the pan briskly from side to side to scatter the seeds up the sides of the pan; set aside.

Combine the rice and oat flours, sugar, remaining butter, and salt in the bowl of the stand mixer and mix on medium speed with the paddle attachment until the mixture is the texture of brown sugar, about a minute. Add the remaining poppy seeds, baking powder, baking soda, xanthan gum, yogurt, eggs, vanilla, and almond extract and beat on medium-high speed for 2 to 3 minutes; the batter should be very smooth and fluffy. Scoop dollops of the batter into the bottom of the pan; don't spread the batter because that will dislodge the seeds. Don't worry if the batter is not quite even; it will level itself in the oven. Bake for 45 to 55 minutes, until a bamboo skewer inserted in the center comes out clean and dry. Set the pan on a rack to cool completely.

When the cake is cool, invert a large plate onto the Bundt pan, and hold it on firmly. Flip the whole business over and remove the pan. Cut into thick slices with a serrated knife to serve. The cake keeps, covered, at room temperature for up to a week.

CHOCOLATE LAYER CAKE

Once upon a time, a good "chocolate layer cake" was a simple yet delicious yellow or white cake filled and frosted with dark, but not too bittersweet, chocolate frosting. This recipe makes that great traditional birthday cake that kids (and everyone else) love; it's a refreshing alternative to the modern über chocolate experience! **SERVES 10 TO 12**

2⅔ cups (400 grams) white rice flour
—OR—
4 cups (400 grams) Thai white rice flour

½ cup (50 grams) gluten-free oat flour

2 cups minus 3 tablespoons (360 grams) sugar

½ pound (2 sticks/225 grams) unsalted butter, very soft

¾ teaspoon salt

2 teaspoons baking powder

1 teaspoon baking soda

½ teaspoon xanthan gum

1 cup plain yogurt (any percent fat) or slightly watered down Greek yogurt

4 large eggs

2 teaspoons pure vanilla extract

Dark Chocolate Frosting (recipe follows)

EQUIPMENT

Two 9-by-2-inch round cake pans

Stand mixer with paddle attachment

Position a rack in the lower third of the oven and preheat the oven to 350°F. Grease the sides of the pans with vegetable oil spray or butter and line the bottoms with parchment paper.

Combine the rice and oat flours, sugar, butter, and salt in the bowl of the stand mixer and mix on medium speed with the paddle attachment until the mixture is the texture of brown sugar, about a minute. Add the baking powder, baking soda, xanthan gum, yogurt, eggs, and vanilla and beat on medium-high speed for 2 to 3 minutes; the batter should be very smooth and fluffy. Scrape into the prepared pans and bake the layers for 25 to 30 minutes, until a toothpick inserted in the center comes out clean and dry. Set the pans on a rack to cool completely.

Slide a thin knife or a small metal spatula around the edges of the cakes to detach them from the pans. Invert one layer onto a serving plate and peel off the parchment liner. Spread about one-quarter of the frosting on top of the layer. Unmold the second layer and peel off the liner. Set it right side up on the frosted layer. Spread another quarter of the frosting in a very thin layer over the top and sides of the cake, just enough to smooth the surface and glue on any crumbs. Chill to set the frosting, about half an hour. Stir the remaining frosting until smooth (warm it slightly if necessary) and spread it over the top and sides of the cake, as luxuriously as you want. Depending on how much frosting you like, there may be some left over. The cake keeps in an airtight container in the refrigerator for up to 5 days. Bring to room temperature before serving.

Dark Chocolate Frosting

MAKES 1 QUART

1 cup heavy cream

17.5 ounces (500 grams) 55% to 62% chocolate, coarsely chopped

½ pound (2 sticks/225 grams) unsalted butter, softened

Put the cream and chocolate in a medium stainless steel bowl. Bring an inch of water to a simmer in a wide skillet. Turn off the heat and set the bowl of chocolate in the water. Let it rest for 15 minutes, gently shaking the bowl several times to submerge the chocolate in the cream. When the chocolate is melted, start whisking at one edge and continue whisking until all of the chocolate is incorporated and the mixture is smooth. Add the butter in chunks and whisk once or twice to break them up; let the mixture rest for 5 minutes before whisking it smooth. Leftovers keep, covered, in the refrigerator for up to 5 days.

CARAMEL APPLE UPSIDE-DOWN CAKE

Tart, flavorful apples, such as Pink Lady, Pippin, Sierra Beauty, Braeburn, Arkansas Black, or Winesap (to name a few), are best with this recipe. Try the pear or nectarine variations; figs are even more unusual and gorgeous. **SERVES 6 TO 8**

FOR THE TOPPING

4 tablespoons (½ stick/55 grams) unsalted butter, very soft

½ cup (100 grams) packed brown sugar

½ teaspoon ground cinnamon

1 large apple

Grated zest and juice of 1 small lemon

FOR THE CAKE

1⅓ cups (200 grams) white rice flour
—OR—
2 cups (200 grams) Thai white rice flour

¼ cup (25 grams) gluten-free oat flour

1 cup minus 2 tablespoons (180 grams) sugar

8 tablespoons (1 stick/115 grams) unsalted butter, very soft

Scant ½ teaspoon salt

1 teaspoon baking powder

½ teaspoon baking soda

¼ teaspoon xanthan gum

½ cup plain yogurt (any percent fat) or slightly watered down Greek yogurt

2 large eggs

1 teaspoon pure vanilla extract

1 pint vanilla ice cream or Whipped Cream (page 343)

EQUIPMENT

9-by-2-inch round cake pan

Stand mixer with paddle attachment

Position a rack in the lowest part of the oven and preheat the oven to 350°F. Use the back of a spoon to smear the 4 tablespoons topping butter all over the bottom of the pan. With the same spoon, spread the brown sugar over the butter (the brown sugar should be in an even layer but does not need to be incorporated into the butter). Sprinkle with the cinnamon. Peel, quarter, and core the apple and cut it into ¼-inch slices. Place in a bowl and toss gently with the lemon zest and juice. Place the apple slices flat in the pan, covering most of the brown sugar layer, and pour the lemon juice from the bowl on top; set aside.

For the cake, combine the rice and oat flours, sugar, butter, and salt in the bowl of the stand mixer and mix on medium speed with the paddle attachment until the mixture is the texture of brown sugar, about a minute. Add the baking powder, baking soda, xanthan gum, yogurt, eggs, and vanilla and beat on medium-high speed for 2 to 3 minutes; the batter should be very smooth and fluffy. Scrape into the prepared pan and spread it evenly. Bake for 40 to 45 minutes, until a toothpick inserted in the center comes out clean.

Let the cake sit for 5 minutes on a rack, then slide a slim knife or small metal spatula around the edge to detach it from the pan. Invert the cake onto a plate to cool. If possible, choose a serving plate that is slightly concave, like a very wide shallow bowl, to accommodate the contour of the cake. If some of the apples stick to the pan, use a spatula to transfer them back into place—the gooey topping will hide all sins here. Serve wedges with ice cream or whipped cream.

(recipe continues)

VARIATIONS

Figgy Upside-Down Cake

Figs are stickier than other fruit: butter the bottom of the pan and line it with parchment paper before smearing it with the topping butter. Substitute 12 small fresh figs for the apple. Stem and halve the figs and arrange them cut side down in the pan. Remove the parchment before serving.

Pear Upside-Down Cake

Substitute 1 large or two small ripe pears for the apple.

Nectarine Upside-Down Cake

Substitute ½ cup (100 grams) granulated sugar for the brown sugar and 1 large nectarine (unpeeled) for the apple.

CINNAMON CRUMB CAKE

Quick breads are especially diverse and strongly regional. Coffee cake is one of the great examples of this—all sorts of quick and yeast-risen breads with and without crumb toppings pass for coffee cake in different parts of the United States. This cinnamon, crumb-topped yellow cake with plenty of butter and eggs also includes yogurt (for a little tanginess) instead of sour cream and a combination of rice and oat flour, which accentuates the flavor of the eggs and butter even better than does wheat flour. After the first day, you will want to toast slices or give them a ten-second warm-up in the microwave. **SERVES 8 TO 12**

1⅔ cups (250 grams) white rice flour
—OR—
2½ cups (250 grams) Thai white rice flour

¼ cup plus 2 tablespoons (40 grams) gluten-free oat flour

1 cup plus 2 tablespoons (240 grams) granulated sugar

10 tablespoons (1¼ sticks/140 grams) unsalted butter, very soft

¾ teaspoon salt

1 cup (100 grams) finely chopped walnuts

½ cup (100 grams) packed brown sugar

1 teaspoon ground cinnamon

1 teaspoon baking powder

½ teaspoon baking soda

¼ teaspoon xanthan gum

¾ cup plain yogurt (any percent fat) or slightly watered down Greek yogurt

2 large eggs

1 teaspoon pure vanilla extract

EQUIPMENT

9-by-13-inch glass baking dish

Stand mixer with paddle attachment

Position the rack in the lower third of the oven and preheat the oven to 350°F. Grease the dish with vegetable oil spray or butter.

Combine the rice and oat flours, granulated sugar, butter, and salt in the bowl of the stand mixer and mix on medium speed with the paddle attachment until the mixture is the texture of brown sugar, about a minute. Remove about 1 cup of the mixture and place it in a small bowl; stir in the walnuts, brown sugar, and cinnamon, and set aside.

Add the baking powder, baking soda, xanthan gum, yogurt, eggs, and vanilla to the mixer bowl and beat on medium-high speed for 2 to 3 minutes; the batter should be very smooth and fluffy. Scrape the batter into the prepared dish and spread it evenly. Scatter the reserved crumb mixture over the batter. Bake for 25 to 30 minutes, until a toothpick inserted in the center comes out clean. Serve warm, cut into 3-inch squares, or let cool. The cake keeps, covered, at room temperature for up to 3 days. Microwave for about 10 seconds to restore freshness.

CARROT SPICE CAKE WITH CREAM CHEESE FROSTING

Rice and oat flours give this moist and otherwise classic American cake a surprising delicacy while letting the spices shine through. A guest once pronounced this the best carrot cake he'd ever eaten. Cool the cake thoroughly and make sure the frosting is soft before topping with Cream Cheese Frosting. **SERVES 12**

1¼ cups flavorless vegetable oil (such as soybean, corn, or safflower)

2 cups (400 grams) sugar

4 large eggs

1½ cups plus 1 tablespoon (240 grams) white rice flour
—OR—
2⅓ cups (240 grams) Thai white rice flour

¾ cup plus 1 tablespoon (80 grams) gluten-free oat flour

1 teaspoon baking soda

2 teaspoons baking powder

2 teaspoons ground cinnamon

½ teaspoon ground nutmeg

¼ teaspoon ground cloves

½ teaspoon salt

3 cups (340 grams) lightly packed shredded peeled carrots (about 4 large carrots)

1 cup (100 grams) coarsely chopped walnuts

Cream Cheese Frosting (recipe follows)

EQUIPMENT

9-by-13-inch glass baking dish

Stand mixer with paddle attachment or handheld mixer

Position a rack in the lower third of the oven and preheat the oven to 350°F. Grease the baking dish with vegetable oil spray or butter.

Combine the oil, sugar, and eggs in the bowl of a stand mixer and beat on medium speed with the paddle attachment until lighter in color, about 2 minutes. Or beat with a handheld mixer on medium-high speed for 3 to 4 minutes.

Add the rice and oat flours, baking soda, baking powder, cinnamon, nutmeg, cloves, salt, carrots, and walnuts and beat on low speed until smooth. Scrape the batter into the prepared dish.

Bake for 30 minutes at 350°F, then reduce the heat to 325°F and bake for 30 minutes longer, or until a toothpick inserted in the center comes out clean. Set the pan on a rack to cool for at least 2 hours before frosting.

To frost, use a spoon to drop dollops of frosting all over the cake, then spread with a small spatula. Cut into 3-inch squares to serve.

The cake keeps, covered, in the refrigerator for up to 5 days.

NOTE: To make this into a layer cake, grease the sides of two 9-by-2-inch round cake pans and line the bottoms with parchment paper. Divide the batter evenly between the pans. Bake for 30 minutes at 350°F and 20 minutes at 325°F. Fill and frost the cake with Cream Cheese Frosting.

(recipe continues)

VARIATION: Sweet Potato Cake

Substitute raw peeled and shredded orange-fleshed sweet potatoes (often called yams) for the carrots.

Cream Cheese Frosting

MAKES 2 CUPS

16 ounces (455 grams) cream cheese

16 tablespoons (2 sticks/230 grams) unsalted butter

3 cups (340 grams) powdered sugar

1 teaspoon pure vanilla extract

Warm the cream cheese and butter in a microwave oven on Low until soft but not melted. Add the powdered sugar and vanilla and beat with a spoon until smooth.

MAYA'S CHOCOLATE FUDGE CAKE WITH MILK CHOCOLATE FROSTING

This is a classic American-style chocolate cake—dense, moist, and dark. The Milk Chocolate Frosting is rich and sweet, with just a touch of fudge-like sugar crystallization; substitute Mocha Mousse Frosting (page 346) for a more sophisticated cake, or make a Red Velvet–Style Cake (see Variations). SERVES 10 TO 12

---

2 cups (400 grams) sugar

1⅓ cups (200 grams) white rice flour
—OR—
2 cups (200 grams) Thai white rice flour

½ cup (50 grams) gluten-free oat flour

⅔ cup (60 grams) natural unsweetened cocoa powder

½ teaspoon salt

¼ teaspoon xanthan gum

1½ teaspoons baking powder

¾ teaspoon baking soda

2 large eggs

1 cup milk

2 teaspoons pure vanilla extract

½ cup flavorless vegetable oil (such as soybean, corn, or safflower)

1 cup boiling water

Milk Chocolate Frosting (recipe follows)

EQUIPMENT

Two 9-by-2-inch or three 8-by-2-inch round cake pans

Stand mixer with paddle attachment

Position a rack in the lower third of the oven and preheat the oven to 350°F. Grease the sides of the cake pans with vegetable oil spray or butter and line the bottoms with parchment paper.

Put the sugar, rice and oat flours, cocoa powder, salt, xanthan gum, baking powder, and baking soda in the bowl of the stand mixer and mix with the paddle attachment until well combined. Add the eggs, milk, and vanilla and beat on medium speed for 2 minutes. Add the oil and beat until smooth. Stir in the hot water until well incorporated. The batter will be thin.

Divide the batter between the prepared pans and bake 8-inch layers for 20 to 25 minutes or 9-inch layers for 30 to 35 minutes, until the cakes pull away slightly from the edges of the pans and a toothpick inserted near the center comes out almost clean. Set the pans on a rack to cool completely before frosting or storing.

When the layers are cool, slide a thin knife or a small metal spatula around the edge of the cakes to detach them from the pans. Invert the layers and peel off the parchment liners. Set one layer on a serving plate. For a two-layer cake, spread one-quarter of the frosting on top of the layer and top with the second layer. For a three-layer cake, spread one-fifth of the frosting on the first layer, top with the second layer, spread with frosting, and top with the third layer. Spread a very thin layer of frosting over the top and sides of the cake just to smooth the surface and glue on any crumbs. Chill to set the frosting, about half an hour. Stir the remaining frosting until smooth (warm it slightly if necessary) and spread it over the top and sides of the cake, as luxuriously as you want. The cake keeps in an airtight container for up to 5 days in the refrigerator. Bring to room temperature before serving.

(recipe continues)

VARIATIONS

Nondairy Fudge Cake

Substitute a total of 2 cups of coconut milk beverage (not Asian coconut milk) or almond milk for the milk and water. Add 1 cup of the liquid cold or at room temperature to replace the milk and 1 cup boiling to replace the water. Fill the layers with cherry preserves and frost with Dark Chocolate Frosting (page 95), made with Asian coconut milk instead of cream and coconut oil instead of butter. (The frosting may need to be chilled briefly—to thicken it—before spreading.)

Red Velvet–Style Cake

Fill and frost the cake with Cream Cheese Frosting (page 102). If you simply must have a red cake, add 1 tablespoon red food coloring to the batter with the vanilla.

Chocolate Fudge Cupcakes

Line a cupcake tin with 12 paper liners and put 3 extra liners in custard cups or ramekins. Make half of the recipe, or any variation, and fill the cups about two-thirds full; there will be exactly enough batter for 15 cupcakes and it will overflow if you put all of the batter in 12. Bake for 20 to 25 minutes, or until a toothpick inserted near the center of a cupcake comes out almost clean.

Milk Chocolate Frosting

MAKES 1 QUART

1 cup heavy cream

17.5 ounces (500 grams) milk chocolate, coarsely chopped

½ pound (2 sticks/225 grams) unsalted butter, softened

Put the cream and chocolate in a medium stainless steel bowl. Bring an inch of water to a simmer in a wide skillet. Turn off the heat and set the bowl of chocolate in the water. Let it rest for 15 minutes, gently shaking the bowl several times to submerge the chocolate in the cream. When the chocolate is melted, start whisking at one edge and continue whisking until all of the chocolate is incorporated and the mixture is smooth. Add the butter in chunks and whisk once or twice to break them up; let the mixture rest for 5 minutes to finish melting the butter before whisking it smooth. Set aside to cool and thicken. Leftovers keep, covered, in the refrigerator for up to 5 days.

CHOCOLATE SHEET CAKE

This is the cake you want to take to the next backyard potluck—frosting is poured over the cake right in the pan, so it can't go anywhere, even if it gets a little warm outside. If you serve the cake chilled, the texture stays nice and soft and moist, and the frosting is easier to slice through neatly. **SERVES 12**

Batter for Maya's Chocolate Fudge Cake (page 103)

1 recipe Milk Chocolate Frosting (page 104)

EQUIPMENT

9-by-13-inch glass baking dish

Position a rack in the lower third of the oven and preheat the oven to 350°F. Grease the baking dish with vegetable oil spray or butter.

Make the batter as directed and scrape it into the prepared dish. Bake for 40 to 45 minutes, until the cake pulls away slightly from the edges of the dish and a toothpick inserted near the center comes out almost clean. Set the dish on a rack to cool completely before frosting. To frost, set the dish on a level surface. If the frosting is too thick to pour, rewarm it for a few seconds at a time in the microwave on Medium (50% power). Pour the frosting over the cake. Let sit at room temperature until the frosting is slightly set, then cover and store for up to 5 days in the refrigerator. Serve chilled (see headnote) or bring to room temperature before serving. To serve, cut into twelve 3-inch squares.

SIMPLE SCONES

These irresistible scones are light and soft inside with a craggy golden brown crust—quite different from the heavy bakery scones you may be accustomed to. Scones rise better if the dough rests in the refrigerator for at least 2 hours and up to 2 days before baking, either as a whole log or sliced. For the best texture, make sure the tops get browned, and don't crowd the scones on the pan. Serve the scones with tea and jam. **MAKES TWELVE 3-INCH SCONES**

1⅓ cups (200 grams) white rice flour (preferably superfine)

—OR—

2 cups (200 grams) Thai white rice flour

½ cup plus 2 tablespoons (60 grams) gluten-free oat flour

¼ cup (50 grams) granulated sugar

¼ teaspoon xanthan gum

1 tablespoon baking powder

½ teaspoon salt

1 cup heavy cream

½ cup plain yogurt (any percent fat) or slightly watered down Greek yogurt

About 2 tablespoons coarse sugar, such as turbinado, for sprinkling

EQUIPMENT

Stand mixer with paddle attachment

Baking sheet, lined with parchment paper

Combine the rice and oat flours, granulated sugar, xanthan gum, baking powder, salt, cream, and yogurt in the bowl of the stand mixer and beat with the paddle attachment for 2 minutes on low speed; the dough will be very stiff. It is important to beat the dough long enough or the scones won't rise well; don't worry about overbeating.

Form the dough into a log 2 inches in diameter, wrap it in plastic, and refrigerate for at least 2 hours or up to 2 days.

Position a rack in the upper third of the oven and preheat the oven to 400°F.

Cut the log into 12 thick slices. Place the slices about 2 inches apart on the lined sheet and sprinkle with coarse sugar. Bake for 20 to 25 minutes, until the scones are browned on both top and bottom. Serve immediately or cool on a rack and toast before serving.

VARIATION: Date-Nut Scones

Beat 1 cup (120 grams) chopped dried pitted dates and 1 cup (100 grams) coarsely chopped walnuts into the dough after mixing.

BUTTER BISCUITS

The combination of rice and oat flours gives these biscuits a subtle complexity; a perfect crisp crust with a soft, pillowy interior; and a pronounced butter flavor. The usual accompaniments are grand here: jam, honey, or sausages and white gravy. Biscuits have the best texture and rise better if the dough rests in the refrigerator for at least 2 hours before baking, either as a whole log or sliced. **MAKES TWELVE 3-INCH BISCUITS**

1⅓ cups (200 grams) white rice flour (preferably superfine)
—OR—
2 cups (200 grams) Thai white rice flour

½ cup plus 2 tablespoons (60 grams) gluten-free oat flour

2 teaspoons sugar

¼ teaspoon xanthan gum

1 tablespoon baking powder

½ teaspoon salt

1 cup heavy cream

½ cup plain yogurt (any percent fat) or slightly watered down Greek yogurt

EQUIPMENT

Stand mixer with paddle attachment

Baking sheet, lined with parchment paper

Combine the rice and oat flours, sugar, xanthan gum, baking powder, salt, cream, and yogurt in the bowl of the stand mixer and beat with the paddle attachment for 2 minutes on low speed; the dough will be very stiff. It is important to beat the dough long enough or the biscuits won't rise well; don't worry about overbeating.

Form the dough into a log 2 inches in diameter, wrap it in plastic, and refrigerate for at least 2 hours or up to 2 days.

Position a rack in the upper third of the oven and preheat the oven to 400°F.

Cut the log into 12 thick slices. Place the slices close together on the pan for soft pull-apart biscuits or 2 inches apart for separate biscuits and bake for 20 to 25 minutes, until browned on top and bottom. Serve immediately or cool on a rack and toast before serving.

YOGURT TART

Supersmooth, rich, slightly tangy yogurt custard is the perfect filling for this oat flour shortbread crust. The elements may sound like the makings of a healthy breakfast, but the results are suave and dressy enough for company. Serve the tart pristinely plain or with a scattering of fresh berries and a drizzle of honey. **SERVES 8 TO 10**

FOR THE CRUST

¾ cup (75 grams) gluten-free oat flour

3 tablespoons (30 grams) white rice flour
—OR—
scant ⅓ cup (30 grams) Thai white rice flour

¼ cup (50 grams) sugar

⅛ teaspoon salt

1/16 teaspoon baking soda

6 tablespoons (85 grams) unsalted butter, cut into chunks and softened

2 tablespoons (30 grams) cream cheese, cut into chunks

½ teaspoon pure vanilla extract

1 large egg yolk, lightly beaten with a pinch of salt

FOR THE FILLING

3 large eggs

¼ cup plus 2 tablespoons sugar (75 grams)

⅛ teaspoon salt

½ teaspoon pure vanilla extract

1½ cups (340 grams/12 ounces) plain Greek yogurt (any percent fat)

EQUIPMENT

9½-inch fluted tart pan with removable bottom

Food processor fitted with the steel blade (optional)

Rimmed baking sheet

Grease the tart pan with vegetable oil spray or butter.

To make the crust by hand, put the oat and rice flours, sugar, salt, and baking soda in a medium bowl and whisk until thoroughly blended. Add the butter, cream cheese, and vanilla. Use a fork or the back of a large spoon to mash and mix the ingredients together until all are blended into a smooth, soft dough.

To make the crust in a food processor, put the oat and rice flours, sugar, salt, and baking soda in the food processor. Pulse to blend. Add the butter, cream cheese, and vanilla. Pulse until the mixture forms a smooth, soft dough. Scrape the bowl and blend in any stray flour at the bottom with your fingers.

Transfer the dough to the tart pan.

The dough may seem much softer than other tart doughs. Use the heel of your hand and then your fingers and/or a small offset spatula to spread the dough all over the bottom of the pan. Press it squarely into the corners with the side of your index finger to prevent extra thickness at the bottom edge, and press it as evenly as possible up the sides of the pan, squaring off along the top edge. Have patience; there is just enough dough (although you may not think so at first). If there is too much dough in one place (or hiding in the corners of the pan), pinch or scrape it off and move it elsewhere. Spread or smear it smooth with the spatula. Here's a final trick for a perfectly even crust: Press a sheet of plastic wrap against the bottom and up the sides of the pan and lay a paper towel on top. Set a straight-sided, flat-bottomed cup on the towel; press and slide the cup all over the bottom and into the corners to smooth and even the surface. Leave the plastic wrap in place. Refrigerate the pan for at least 2 hours, but preferably overnight and up to 3 days.

(recipe continues)

Position a rack in the lower third of the oven and preheat the oven to 325°F.

Peel the plastic wrap from the crust. Set the pan on the baking sheet and bake the crust for 35 to 40 minutes, checking after 15 to 20 minutes. If the crust has puffed up on the bottom, press it back down carefully with the back of a fork. Continue baking until the crust is golden brown and has pulled away from the sides of the pan.

While the crust is baking, make the filling: Whisk the eggs with the sugar, salt, and vanilla in a medium bowl. Whisk in the yogurt.

When the crust is ready, remove it from the oven and turn the temperature down to 300°F. Brush the bottom of the crust with a thin layer of the beaten egg yolk to moisture-proof it. (Discard any remaining egg wash or save it for another use.) Return the crust to the oven for 2 minutes to set the yolk.

Pour the filling into the hot crust and spread it evenly. Return the tart to the oven and bake until the filling is set around the edges but, when the pan is nudged, quivers like very soft Jell-O in the center, 15 to 20 minutes. Check often in the last few minutes, as overbaking will destroy the silky-smooth texture of the filling. Cool the tart completely on a rack. Refrigerate if not serving within 3 hours. To serve, remove the sides of the pan and transfer the tart (with or without the pan bottom) to a platter. The tart is best on the day it's made, but leftovers may be kept in a covered container in the refrigerator for a day or two.

BLACKBERRY GALETTE

Use juicy blackberries or raspberries, olallieberries, or boysenberries to make a delicately flaky free-form tart. SERVES 6

¼ recipe rugelach dough (see page 121)

¼ teaspoon white rice flour, plus more for rolling

3 tablespoons (35 grams) sugar

3 cups (385 grams) blackberries

Cinnamon sugar: 1 tablespoon sugar mixed with ¼ teaspoon ground cinnamon

EQUIPMENT

Rolling pin

Baking sheet

Make the dough as directed and shape it into a round patty. Chill as directed.

Position a rack in the lower third of the oven and preheat the oven to 400°F.

Remove the dough from the refrigerator. Let it stand for a few minutes until just pliable enough to roll, but not too soft. Use a rolling pin to roll the dough between sheets of wax paper or parchment into a 12-inch circle a scant ⅛ inch thick. Slide the dough and paper onto a baking sheet and refrigerate for 10 to 15 minutes so it will be easier to handle. (Don't bother to make shelf space in the fridge; just slip the pan on top of whatever is in there!) Remove the pan from the refrigerator and slide the dough out onto the counter. Peel off the top sheet of paper and set it clean side up on the pan. Dust the dough very lightly with rice flour. Lift the paper under the dough and flip the dough over onto the parchment-lined pan. Peel off the paper.

Mix the sugar and ¼ teaspoon rice flour and toss with the berries in a medium bowl. Pour the berries and sugar in the center of the dough, leaving a 2½-inch border all around. Fold the edges of the dough over the berries, pleating it as you go. Don't worry if the dough breaks or cracks a little. Bush off any excess rice flour. Sprinkle the edges of the dough with pinches of cinnamon sugar.

Bake for 25 to 30 minutes, until the dough is golden to deep golden brown on top and deep golden brown on the bottom (lift the parchment a little to peer underneath). Set the pan on a rack to cool. Slide the galette onto a serving platter and serve warm or at room temperature. The galette is best on the day it is made.

PEACH CRUMBLE

The contrast between crunchy topping and juicy fruit is enticing, but it turns out (after numerous overly gooey trials) that it is difficult to achieve with any fruit softer than an apple. Preparing the topping and fruit separately may seem like an annoying extra step, but is actually easier and more foolproof, since you don't need to coordinate the doneness of the topping and filling—just unite them when the peaches are done. **SERVES 6 TO 8**

FOR THE FRUIT

About 2 pounds (900 grams) peaches, unpeeled, cut into ¾-inch chunks, to make 5 cups

2 tablespoons lemon juice

¼ cup (50 grams) sugar

1 tablespoon white rice flour

FOR THE TOPPING

⅔ cup (100 grams) white rice flour
—OR—
1 cup (100 grams) Thai white rice flour

¼ cup (25 grams) gluten-free oat flour

¼ teaspoon salt

1 cup (85 grams) sliced natural (unskinned) almonds

1 cup (200 grams) packed light brown sugar

¼ teaspoon ground cloves

1¼ teaspoons ground cinnamon

6 tablespoons (85 grams) unsalted butter, melted

EQUIPMENT

2-quart baking dish, 2 to 3 inches deep

Baking sheet, lined with parchment paper

Position oven racks in the lower and upper thirds of the oven and preheat the oven to 400°F.

For the fruit, combine the peaches, lemon juice, sugar, and rice flour in the baking dish. Bake for 15 minutes on the lower rack, stir, and bake for 30 to 35 more minutes, or until the juices are bubbling at the center of the dish.

Make the topping while the peaches are baking: Combine the rice and oat flours, salt, nuts, brown sugar, cloves, cinnamon, and butter in a medium bowl. Stir until well combined. Transfer the mixture to the lined baking sheet and spread it evenly about ½ inch thick. Bake on the upper rack for 8 to 10 minutes, until well browned. Set aside. As soon as the peaches are ready, use a spatula to cover them with large pieces of topping. Cool for at least 20 minutes and serve warm or at room temperature. The crumble can be kept at room temperature, covered with a paper towel, for up to 1 day or covered tightly with plastic wrap and refrigerated for up to 3 days.

VARIATIONS

Nectarine, Apricot, Plum, or Cherry Crumble

Substitute 5 cups cubed nectarines, apricots, or plums or 5 cups (755 grams) pitted cherries for the peaches. Add 2 tablespoons more sugar to the fruit.

Blueberry, Blackberry, or Huckleberry Crumble

Substitute 5 cups (700 grams) fresh or frozen (not defrosted) berries for the peaches. (Mixed blueberries and huckleberries are especially good!) Add 2 tablespoons more sugar to the fruit. Substitute 1 cup (100 grams) finely chopped walnuts or pecans for the almonds.

APPLE CRUMBLE

To peel or not to peel the apples! I often make crumbles and crisps with unpeeled apples, because I like the texture and the extra flavor that comes from the skins, but you can have it your way. Tender apples with tender skins need the skins so they won't fall apart, but denser apples with tougher skins may be a little too chewy (for some tastes). If your apples are particularly sweet, you might want to add the extra tablespoon of lemon juice. **SERVES 6 TO 8**

FOR THE FRUIT

About 2 pounds (900 grams) apples, peeled or unpeeled, cored and cut into ¾-inch chunks to make 5 cups

2 to 3 tablespoons lemon juice (see headnote)

¼ cup water

2 tablespoons (25 grams) sugar

1 tablespoon white rice flour

FOR THE TOPPING

½ cup (100 grams) granulated sugar

⅔ cup (100 grams) white rice flour
—OR—
1 cup (100 grams) Thai white rice flour

¼ cup (25 grams) gluten-free oat flour

¼ teaspoon salt

1 cup (100 grams) finely chopped walnuts

½ cup (100 grams) packed light brown sugar

¼ teaspoon ground nutmeg

1¼ teaspoons ground cinnamon

6 tablespoons (85 grams) unsalted butter, melted

EQUIPMENT

2-quart baking dish, 2 to 3 inches deep

Rimmed baking sheet

Position a rack in the lower third of the oven and preheat the oven to 400°F.

For the fruit, combine the apples, lemon juice, water, sugar, and rice flour in the baking dish. Place the dish on the rimmed baking sheet and bake for 15 minutes, stir, and bake for 15 to 20 more minutes, until the juices are bubbling at the edge of the dish.

Make the topping while the apples are baking: Combine the granulated sugar, rice and oat flours, salt, walnuts, brown sugar, nutmeg, cinnamon, and melted butter in a medium bowl. Stir until well combined. When the apples are ready, spoon the topping over them. Bake for 15 to 20 more minutes, until the topping is browned and the apples are tender and bubbling in the center of the dish. Cool for at least 20 minutes and serve warm or at room temperature. The crumble can be kept at room temperature, covered with a paper towel, for up to 1 day or covered tightly with plastic wrap and refrigerated for up to 3 days.

VARIATION: Pear Crumble

Substitute 5 cups hard-ripe pears, such as Bartlett (Williams) or Bosc, cut into cubes, for the apples. If you have tender pears, use the recipe for Peach Crumble (page 115)—it's better for softer fruits—and substitute 5 cups cubed tender pears for the peaches.

BUTTERY APPLE COBBLER

Cobbler is easy fall comfort food, and especially satisfying if you've been out picking the apples. Cream instead of butter makes the biscuit topping quick to mix, and—surprisingly—extra fresh and buttery tasting. SERVES 6 TO 8

FOR THE FRUIT

About 2 pounds (900 grams) apples, peeled, quartered, cored, and roughly sliced about ⅓ inch thick, to make 5 cups

2 tablespoons lemon juice

1 tablespoon white rice flour

½ cup (100 grams) sugar

FOR THE BISCUIT TOPPING

⅔ cup (100 grams) white rice flour (preferably superfine)
—OR—
1 cup (100 grams) Thai white rice flour

¼ cup plus 1 tablespoon (30 grams) gluten-free oat flour

⅛ teaspoon xanthan gum

1 tablespoon granulated sugar

1½ teaspoons baking powder

¼ teaspoon salt

½ cup heavy cream

¼ cup plain yogurt (any percent fat) or slightly watered down Greek yogurt

1 tablespoon coarse sugar, such as turbinado, for sprinkling

EQUIPMENT

2-quart baking dish, 2 to 3 inches deep

Rimmed baking sheet

Stand mixer with paddle attachment

Position a rack in the lower third of the oven and preheat the oven to 400°F.

For the fruit, combine the apples, lemon juice, rice flour, and sugar in the baking dish. Place on the rimmed baking sheet and bake for 15 minutes, stir, and bake for 15 to 20 more minutes, or until the juices are bubbling at the edge of the dish.

Mix the biscuit dough while the apples are baking: Combine the rice and oat flours, xanthan gum, granulated sugar, baking powder, salt, cream, and yogurt in the bowl of the stand mixer, and beat with the paddle attachment for 2 minutes on low speed; the dough will be very stiff. It is important to beat the dough long enough or the biscuits won't rise well; don't worry about overbeating.

When the apples are ready, spoon dollops of biscuit dough over them (don't cover the apples completely) and sprinkle with coarse sugar. Bake for 15 to 20 minutes, or until the topping is browned and the apples are tender and bubbling in the center. Cool for at least 20 minutes and serve warm or at room temperature.
The cobbler can be kept at room temperature for up to 1 day, covered with a paper towel, or covered tightly with plastic wrap and refrigerated for up to 3 days.

VARIATION: Pear and Rhubarb Cobbler

Substitute 5 cups of mixed sliced pears and fresh or frozen rhubarb slices for the apples and add 2 tablespoons sugar to the filling.

NEW CLASSIC BLONDIES

Oat flour brings its own natural butterscotch flavor to this childhood treat.

MAKES SIXTEEN 2-INCH BLONDIES

⅔ cup (60 grams) gluten-free oat flour

⅓ cup plus 1 tablespoon (60 grams) white rice flour
—OR—
½ cup plus 1 tablespoon (60 grams) Thai white rice flour

3 tablespoons (30 grams) potato starch

¼ teaspoon baking soda

¼ teaspoon xanthan gum

Rounded ¼ teaspoon salt

8 tablespoons (1 stick/115 grams) unsalted butter, melted and kept warm

⅓ cup (65 grams) packed light brown sugar

⅓ cup (65 grams) granulated sugar

1 large egg

¾ teaspoon pure vanilla extract

¾ cup (75 grams) walnut pieces

¾ cup (130 grams) semisweet or bittersweet chocolate chips

EQUIPMENT

8-inch square pan, bottom and all four sides lined with parchment paper or foil

Position a rack in the lower third of the oven and preheat the oven to 350°F.

In a small bowl, whisk the oat and rice flours, potato starch, baking soda, xanthan gum, and salt until thoroughly blended. Set aside.

Mix the butter and brown and granulated sugars in a medium bowl. Use a spatula or a wooden spoon to beat in the egg and vanilla. Stir in the flour mixture just until moistened, then beat about 20 strokes to aerate the batter slightly. Stir in half of the walnuts and half of the chocolate chips. Spread the batter in the lined pan. Scatter the remaining walnuts and chocolate chips evenly over the top.

Bake for 20 to 25 minutes, or until the nuts look toasted, the top is golden brown, and the edges have pulled away from the sides of the pan. Cool in the pan on a rack. Lift the ends of the parchment or foil and transfer to a cutting board. Use a long sharp knife to cut into 16 squares. The blondies may be stored in an airtight container for 3 to 4 days.

OAT AND ALMOND TUILES

Oat flour adds sensational buttery toffee notes to these otherwise classic almond wafers—and to the equally marvelous coconut version that follows. If you want curved or shaped tuiles, see page 69. **MAKES ABOUT FORTY 3-INCH COOKIES**

---

4 tablespoons (½ stick/55 grams) unsalted butter, melted, plus more for greasing the foil

2 large egg whites

2 teaspoons water

½ cup (100 grams) sugar

¼ cup plus 1 tablespoon (30 grams) gluten-free oat flour

¼ teaspoon pure almond extract

Scant ½ teaspoon salt

⅔ cup (70 grams) sliced almonds

EQUIPMENT

2 baking sheets

Silicone baking mats (optional)

Rolling pin or small cups for shaping (optional); see page 69

Line the baking sheets with regular foil (dull side facing up) or silicone mats and grease the foil or silicone lightly but thoroughly, or line the sheets with nonstick foil, nonstick side up (see Note, page 69).

In a medium bowl, mix the egg whites with the water, sugar, oat flour, almond extract, and salt until well blended. Stir in the butter and almonds. Cover the bowl and let the batter rest for several hours or overnight in the refrigerator to hydrate the flour.

Position racks in the upper and lower thirds of the oven (or one rack in the center if you are baking only one sheet at a time) and preheat the oven to 325°F.

Stir the batter. Drop level teaspoons 2 inches apart on the lined baking sheets. Use the back of the spoon to smear the batter into 2½-inch rounds—tuiles baked on greased foil will spread a little; those baked on silicone or nonstick foil will not. Bake, watching carefully, for 12 to 15 minutes, rotating the sheets from front to back and top to bottom about halfway through the baking time, until the tuiles are mostly deep golden brown. If the cookies are not baked enough, they will not be completely crisp when cool, nor will they come off the foil easily.

Slide the foil sheets onto racks and let the cookies cool completely before removing them. Or, if using silicone mats, transfer the hot cookies to a rack to cool.

To retain crispness, put the cookies in an airtight container as soon as they are cool. They may be stored airtight for at least 1 month.

VARIATION: Oat and Coconut Tuiles

Substitute ⅓ cup (25 grams) unsweetened dried shredded coconut for the almonds.

APRICOT WALNUT RUGELACH

Part of a great eastern European Jewish baking tradition, rugelach, though considered cookies, are really miniature pastries: flaky cream cheese dough rolled up around cinnamon sugar, jam, currants, and nuts (the traditional filling), or fanciful variations that may include bits of chocolate or cacao nibs or whatever good things a baker might have on hand. Not surprisingly, the oat flour is a perfect flavor partner for the fruits, nuts, and spices in this recipe and all of the variations. **MAKES 48 COOKIES**

FOR THE DOUGH

½ pound (2 sticks/225 grams) unsalted butter, cold

1⅓ cups (200 grams) white rice flour

—OR—

2 cups (200 grams) Thai white rice flour, plus more for rolling

1¾ cups (175 grams) gluten-free oat flour

8 ounces (225 grams) cream cheese

2 tablespoons (25 grams) sugar

1 teaspoon xanthan gum

½ teaspoon baking soda

¼ teaspoon salt

¼ cup water

FOR THE FILLING

¾ cup (235 grams) thick apricot jam or preserves

1 teaspoon ground cinnamon

1 cup (225 grams) finely chopped walnuts

½ cup (65 grams) chopped dried apricots

Salt

Cinnamon sugar: 2 tablespoons sugar mixed with ½ teaspoon ground cinnamon

To make the dough, using the largest holes on a box (or other) grater, grate the butter onto a plate lined with wax paper. Refrigerate until needed.

Combine the rice and oat flours in a bowl and mix thoroughly with a whisk.

In the bowl of a stand mixer with the paddle attachment or in a large bowl with a handheld mixer, mix the cream cheese, sugar, xanthan gum, baking soda, salt, and water for about 2 minutes on medium speed. The mixture will look wet and stretchy.

Add the flour mixture and beat on low speed until the mixture resembles coarse bread crumbs (it will not be smooth). Sprinkle the shredded butter into the bowl and mix on low speed to break the butter shreds into bits and distribute them. The mixture will resemble loose crumbs, sticking together only when pinched. If necessary, sprinkle and mix in another tablespoon of water. Do not try to form a cohesive dough. Divide the mixture into quarters. Dump one-quarter in the center of a sheet of plastic wrap. Bring the sides of the wrap up around the mixture on all sides, pressing firmly to form a 5-inch square patty. Wrap the patty tightly. Repeat with the remaining 3 portions of dough. Refrigerate the patties until firm, at least 2 hours and up to 3 days.

Position racks in the upper and lower thirds of the oven and preheat the oven to 350°F.

Remove 1 piece of dough from the refrigerator. If necessary, let it stand until pliable enough to roll, but not too soft. Roll between sheets of wax paper or parchment paper into a 12-inch round a scant ⅛ inch thick. Slide the wax paper and dough onto a baking sheet and refrigerate. Repeat with the remaining pieces of dough,

EQUIPMENT

Box grater

Stand mixer with paddle attachment or handheld mixer

Rolling pin

Baking sheets, lined with parchment paper or foil

stacking them in the refrigerator. Chill the dough for at least 15 minutes. Meanwhile, pulse the preserves in a food processor if there are large pieces of fruit and stir in the cinnamon.

Set one piece of dough on the counter and peel off the top sheet of wax paper and turn it clean side up on the counter or a cutting board. Dust the dough very lightly with a little rice flour and flip it onto the paper and peel off the second sheet. Spread one-quarter of the preserves over the dough and sprinkle with one-quarter of the walnuts, one-quarter of the apricots, and a tiny pinch of salt. Cut the dough like a pie into 12 equal wedges. Roll the wide outside edge up around the filling toward the point, brushing off any excess rice flour as you go. Place the roll, with the dough point underneath to prevent it from unrolling, on a lined baking sheet. Repeat with the remaining wedges, placing cookies 1½ inches apart. If at times the dough becomes too soft to roll, return it to the refrigerator to firm up. Fill, cut, and roll the remaining pieces of dough. Sprinkle with cinnamon sugar.

Bake for 18 to 20 minutes, until the cookies are golden brown at the edges and deep brown on the bottom. Rotate the sheets from top to bottom and from front to back halfway through the baking time to ensure even baking. Set the pans or just the liners on racks to cool. Let the rugelach cool completely before stacking or storing. The rugelach are always most exquisite on the day they are baked, but they remain delicious, stored in an airtight container, for about 5 days.

VARIATIONS

Blueberry Walnut Rugelach

Omit the apricot preserves. Mix the walnuts and cinnamon with 2 tablespoons (25 grams) granulated sugar, ½ cup (100 grams) packed brown sugar, and ½ cup dried blueberries (or substitute currants). Sprinkle one-quarter of the mixture over each round of dough, and roll over the filling with a rolling pin to press it gently into the dough before cutting into wedges. Sprinkle with salt, roll up, and bake as directed.

Chocolate-Hazelnut Rugelach

Combine ½ cup (100 grams) granulated sugar, 1 teaspoon pure vanilla extract, 1 cup (225 grams) finely chopped toasted and skinned hazelnuts, and 1 cup (170 grams) miniature chocolate chips. Use in place of the apricot filling as follows: Sprinkle one-quarter of the mixture over each round of dough, and roll over the filling with a rolling pin to press it gently into the dough before cutting into wedges. Sprinkle with salt, roll up, and bake as directed.

Cacao Nib Rugelach

Combine 2 tablespoons (25 grams) granulated sugar, ½ cup (100 grams) packed light brown sugar, 1 teaspoon ground cinnamon, ½ cup (70 grams) dried currants, and ½ cup (55 grams) roughly chopped cacao nibs for the walnuts. Use in place of the apricot filling as follows: Sprinkle one-quarter of the mixture over each round of dough, and roll over the filling with a rolling pin to press it gently into the dough before cutting into wedges. Sprinkle with salt, roll up, and bake as directed.

DOUBLE OATMEAL COOKIES

These oatmeal cookies—made with oats *and* oat flour—are truly for oat lovers! You can adjust the chewiness of your cookie by mixing the batter more or mixing it less. Baking them on foil (instead of parchment) or directly on a greased pan produces some lovely crunchy edges on these otherwise chewy cookies. **MAKES ABOUT THIRTY-TWO 3½-INCH COOKIES**

1¼ cups (125 grams) gluten-free oat flour

2 cups (190 grams) rolled oats

½ teaspoon salt

1 teaspoon baking soda

¾ teaspoon xanthan gum

½ pound (2 sticks/225 grams) unsalted butter, melted

¾ cup (150 grams) granulated sugar

¾ cup (150 grams) packed light brown sugar

1 teaspoon ground cinnamon

¼ teaspoon freshly grated nutmeg

1 teaspoon pure vanilla extract

2 large eggs

Generous 1 cup (115 grams) coarsely chopped or broken walnut pieces

1 cup (140 grams) raisins

EQUIPMENT

Baking sheets, lined with foil (dull side up) or greased

Combine the oat flour, rolled oats, salt, baking soda, and xanthan gum in a medium bowl and whisk until blended.

In a large bowl, mix the butter, sugars, cinnamon, nutmeg, and vanilla. Whisk in the eggs. Stir in the oat mixture and mix the batter briskly with a spatula for about 1 minute (to activate the binding power of the xanthan gum—the more you mix, the chewier and less crunchy the cookies will be). Stir in the nuts and raisins. Let the dough stand for at least 1 but preferably 2 hours or (better still) cover and refrigerate overnight.

Position racks in the upper and lower thirds of the oven and preheat the oven to 325°F.

Scoop 2 tablespoons of the dough per cookie and place 2 inches apart on the lined or greased baking sheets. Bake for 16 to 20 minutes, until the cookies are deep golden brown. Rotate the sheets from top to bottom and from front to back halfway through the baking time to ensure even baking. For lined pans, set the pans or just the liners on racks to cool; for unlined pans, use a metal spatula to transfer the cookies to racks. Cool the cookies completely before stacking or storing. They may be kept in an airtight container for several days.

CLASSIC GINGER COOKIES

The flavor of oat flour is so perfect with the ginger and other warm spices in these easy one-bowl cookies, I can't believe I didn't think of using this combination sooner. These new ginger cookies may become your favorites! **MAKES ABOUT 50 COOKIES**

½ cup (100 grams) granulated sugar

⅓ cup (65 grams) packed brown sugar

2½ teaspoons ground ginger

1½ teaspoons ground cinnamon

½ teaspoon ground allspice

2 teaspoons baking soda

¼ teaspoon salt

¼ cup (85 grams) unsulfured mild or full-flavored molasses (not blackstrap)

8 tablespoons (1 stick/115 grams) unsalted butter, melted and kept lukewarm

2 large egg whites

2¾ cups (275 grams) gluten-free oat flour

¾ cup (115 grams/4 ounces) ginger chips or crystallized ginger, cut into ¼-inch dice, shaken in a coarse strainer to remove loose sugar

About ½ cup (100 grams) granulated or coarse sugar, such as turbinado, for rolling

EQUIPMENT

Baking sheets, lined with parchment paper or unlined and ungreased

If you are baking the cookies right away, position racks in the upper and lower thirds of the oven and preheat the oven to 350°F.

In a medium bowl, mix the ½ cup granulated sugar, the brown sugar, ground ginger, cinnamon, allspice, baking soda, salt, molasses, butter, and egg whites until blended. Stir in the flour.

When the flour is no longer visible, beat the dough briskly with a spatula or a wooden spoon, about 40 strokes, to aerate it slightly. Stir in the ginger chips. The dough will be very soft. If possible, chill it for an hour or two to firm it up, or (better still) cover and refrigerate it overnight for the best flavor and texture.

Form the dough into 1-inch balls (15 grams each). Roll the balls in granulated or coarse sugar and place them 2 inches apart on the lined or ungreased baking sheets. Bake for 10 to 12 minutes, or until the cookies puff up and crack on the surface and then begin to deflate in the oven. Rotate the sheets from top to bottom and from back to front halfway through the baking time to ensure even baking. For chewier cookies, remove them from the oven when at least half or more of the cookies have begun to deflate; for crunchier edges with chewy centers, bake for a minute or so longer.

For lined pans, set the pans or just the liners on racks to cool; for unlined pans, use a metal spatula to transfer the cookies to racks. Cool the cookies completely before storing. They may be kept in an airtight container for several days.

VARIATION: Molasses Spice Cookies

Substitute ¾ cup plus 2 tablespoons (175 grams) light or dark brown sugar for the granulated and brown sugar in the dough. Substitute ½ teaspoon ground cloves for the allspice. Substitute 1 large egg for the egg whites. Omit the ginger chips. Makes about 40 cookies.

CHOCOLATE CHIP COOKIES

Oat and brown rice flours give these cookies extra butterscotch flavor—and a beautiful butterscotch color as well. You will love the buttery, crunchy edges and delicate cakey interiors. Most guests are likely to think you've simply gone and made some really good chocolate chip cookies; they'll be hard put to identify the particulars! And if you know people who cannot eat wheat but long for a remembered treat, these are for them as well. Resting the dough overnight—or for at least a couple of hours—makes these cookies especially good. **MAKES ABOUT SIXTY 3½-INCH COOKIES**

1¼ cups (125 grams) gluten-free oat flour

1 cup (135 grams) brown rice flour

¼ cup plus 2 tablespoons (65 grams) potato starch

½ teaspoon salt

½ teaspoon baking soda

¾ teaspoon xanthan gum

½ pound (2 sticks/225 grams) unsalted butter, melted

¾ cup (150 grams) granulated sugar

¾ cup (150 grams) packed dark brown sugar

1 teaspoon pure vanilla extract

2 large eggs

2 cups (340 grams) chocolate chips or chunks or hand-chopped chocolate

1 cup (100 grams) walnuts or pecans, coarsely chopped

EQUIPMENT

Baking sheets, lined with foil (dull side up) or greased

Combine the flours, potato starch, salt, baking soda, and xanthan gum in a medium bowl and mix thoroughly with a whisk.

In a large bowl, mix the melted butter, sugars, and vanilla. Whisk in the eggs. Stir in the flour mixture. With a rubber spatula, mix the batter briskly for about 45 seconds (to activate the binding power of the xanthan gum—the more you mix, the chewier and less crunchy the cookies will be). Stir in the chocolate chips and nuts. If possible, let the dough stand for 1 to 2 hours or (better still) cover and refrigerate overnight.

Position racks in the upper and lower thirds of the oven and preheat the oven to 375°F.

Scoop 2 tablespoons of dough per cookie and place 2 inches apart on the lined or greased baking sheets. Bake until the cookies are golden brown, 12 to 14 minutes. Rotate the pans from top to bottom and from front to back halfway through the baking time to ensure even baking. For lined pans, set the pans or just the liners on racks to cool; for unlined pans, use a metal spatula to transfer the cookies to racks. Cool the cookies completely before stacking or storing. They may be kept in an airtight container for several days.

VARIATIONS

Pecan Spice Cookies

Add 2 teaspoons ground cinnamon, ½ teaspoon ground nutmeg, ½ teaspoon ground cloves, and ½ teaspoon ground ginger with the sugar. Omit the chocolate chips and use 2 cups (200 grams) lightly toasted pecan pieces for the nuts.

Nibby Nut and Raisin Cookies

Omit the chocolate chips. Add 1 cup (140 grams) raisins and ⅔ cup (75 grams) roasted cacao nibs with the walnuts.

OAT SABLÉS

These are splendid plain cookies—even more buttery and flavorful than traditional butter cookies made with all-purpose wheat flour. They have a perfect melt-in-your-mouth sandy texture and a gentle nuance of oat flavor. They are easy slice-and-bake cookies, but you can also roll out the dough and cut shapes for a child's party or holiday cookie decorating (see page 131). As plain and good as they are, feel free to embellish with chopped nuts or cocoa nibs, or try any of the other six "modern" combinations that follow. **MAKES ABOUT THIRTY-SIX 2-INCH COOKIES**

1¼ cups plus 2 tablespoons (140 grams) gluten-free oat flour

¼ cup plus 2 tablespoons (55 grams) white rice flour

—OR—

½ cup plus 1 tablespoon (55 grams) Thai white rice flour

¼ teaspoon salt

⅛ teaspoon baking soda

⅔ cup (130 grams) sugar

¼ cup (60 grams) cream cheese, cut into chunks

12 tablespoons (1½ sticks/170 grams) unsalted butter, softened and cut into chunks

1 teaspoon pure vanilla extract

EQUIPMENT

Food processor fitted with the steel blade (optional)

Baking sheets, lined with parchment paper

To make the dough by hand, put the oat and rice flours, salt, and baking soda in a large bowl and whisk until thoroughly blended. Add the sugar, cream cheese, butter, and vanilla. Use a fork or the back of a large spoon to mash and mix the ingredients together until all are blended into a smooth, soft dough.

To make the dough in a food processor, put the oat and rice flours, salt, and baking soda in the food processor. Pulse to blend. Add the sugar, cream cheese, butter, and vanilla. Pulse until the mixture forms a smooth, soft dough. Scrape the bowl and blend in any stray flour at the bottom with your fingers.

Scrape the dough onto a sheet of wax paper and form it into a 10-inch log about 1¾ inches in diameter. Wrap it tightly in the wax paper and refrigerate for at least 2 hours, but preferably longer or overnight. The dough may be frozen for up to 3 months.

Position racks in the upper and lower thirds of the oven and preheat the oven to 325°F.

Use a sharp knife to cut the cold dough log into ¼-inch-thick slices. Place the cookies at least 1½ inches apart on the lined baking sheets. Bake for 15 to 20 minutes, until the cookies are golden brown at the edges and well browned on the bottom. Rotate the pans from top to bottom and from front to back halfway through the baking time to ensure even baking.

Set the pans or just the liners on racks to cool. Cool the cookies completely before stacking or storing. They may be kept in an airtight container for at least 2 weeks.

VARIATIONS

Nutty Oat Sablés

Add ½ cup (70 grams) of any raw or toasted nuts to the dough: If mixing by hand, add them to the dough at the end. If using a food processor, add nuts whole with the dry ingredients and pulse until they are the desired consistency.

Nibby Oat Sablés

Add a generous ¼ cup (35 grams) roasted cacao nibs to the dough with the butter.

Grapefruit and Basil Sablés

Omit the vanilla and add 1½ teaspoons finely grated grapefruit zest and 1½ teaspoons finely chopped fresh basil leaves with the butter.

Orange-Saffron Sablés

Omit the vanilla and add 1½ teaspoons finely grated orange zest and ¾ teaspoon loosely packed crumbled saffron threads with the butter.

Orange Sablés with Ancho Chile

Omit the vanilla and add 1½ teaspoons finely grated orange zest and 1½ teaspoons crumbled or powdered dried ancho chile with the butter.

Chai Sablés

Add 2 teaspoons pulverized chai (from a package of loose chai or from the contents of 2 to 3 chai tea bags) with the butter.

Lime and Mint Sablés

For a little mojito flavor, omit the vanilla and add 1½ teaspoons finely grated lime zest and 1½ teaspoons finely chopped fresh mint leaves with the butter.

Spicy Basil Sablés

Omit the vanilla and add 1½ teaspoons ground cinnamon and 1½ teaspoons finely chopped fresh basil leaves with the butter.

CUTOUT COOKIES

When you need simple tasty cutout cookies for a child's party or a themed occasion, these are a sure thing. To decorate, simply sprinkle them with colored sugar before baking, or bake and cool the cookies first and then ice them—or pipe melted chocolate on them—before affixing sugars, sprinkles, miniature candies, and so on. **MAKES ABOUT THIRTY-SIX 2½-INCH COOKIES**

Oat Sablés dough (page 128)

Gluten-free oat flour for cutting out the cookies

Colored sugars for sprinkling (optional)

EQUIPMENT

Rolling pin

Cookie cutters

Baking sheets, lined with parchment paper

NOTE: Instead of chilling the dough first and rolling it later, you can roll the freshly made dough gently (it will be very soft) between sheets of wax paper immediately, and then stack and refrigerate the rolled-out dough on a baking sheet for at least 2 hours or until needed.

Mix the dough for Oat Sablés as directed and form it into 2 flat patties. Wrap and refrigerate the dough for at least 2 hours, but preferably overnight (see Note).

Position racks in the upper and lower thirds of the oven and preheat the oven to 325°F.

Remove 1 patty from the refrigerator and let it sit at room temperature briefly, until supple enough to roll but still quite firm. It will continue to soften as you work. Roll the dough between two pieces of wax paper, or between heavy plastic sheets cut from a plastic bag, to a thickness of ⅛ inch. Turn the dough over once or twice while rolling it out to check for deep wrinkles; if necessary, peel off and smooth the paper or plastic over the dough before continuing to roll it. When the dough is thin enough, peel off the top sheet of paper or plastic and keep it in front of you. (If the dough is sticky, dust it with a little oat flour.) Invert the dough onto the sheet in front of you and peel off the second sheet.

Cut cookie shapes as close together as possible to minimize scraps, dipping the edges of cookie cutters in oat flour as necessary to prevent sticking. Use the point of a paring knife to lift and remove scraps as you transfer cookies to the baking sheets. Place the cookies ½ inch apart. If the dough gets too soft at any time—while rolling, cutting, removing scraps between cookies, or transferring cookies—slide a baking sheet underneath the paper or plastic and refrigerate the dough for a few minutes until firm. Repeat with the second piece of dough.

Press all of the dough scraps together gently and reroll them as necessary. (Don't worry that rerolling scraps will produce tough cookies.) Sprinkle the cookies with colored sugars if desired, and pat to adhere.

(recipe continues)

Bake for 8 to 12 minutes, until golden brown at the edges but deep brown on the bottom, rotating the baking sheets from top to bottom and from front to back halfway through the baking time to ensure even baking. Repeat until all the cookies are baked.

Set the pans or just the liners on racks to cool. Cool completely before icing, stacking, or storing. The cookies may be kept in an airtight container for at least 2 weeks.

NUTTY THUMBPRINT COOKIES

These thumbprint cookies are actually a filled version of ultra-tender, not-too-sweet Russian tea cakes, which are remarkably similar to Mexican wedding cakes or those divine crescent-shaped Austrian cookies made with almonds or hazelnuts instead of walnuts or pecans. The combination of rice and oat flours makes these especially tender and flavorful. To make tea cakes or wedding cakes instead of thumbprints, skip the "poking" and filling steps and just dust the cookies with powdered sugar. **MAKES THIRTY-SIX TO FORTY 1½-INCH COOKIES**

1½ cups (150 grams) walnuts or pecans

¼ cup plus 2 tablespoons (55 grams) white rice flour
—OR—
½ cup plus 1 tablespoon (55 grams) Thai white rice flour

1¼ cups plus 2 tablespoons (140 grams) gluten-free oat flour

¼ teaspoon salt

⅛ teaspoon baking soda

⅓ cup (65 grams) granulated sugar

¼ cup (60 grams) cream cheese, cold, cut into chunks

12 tablespoons (1½ sticks/170 grams) unsalted butter, slightly softened and cut into chunks

1 teaspoon pure vanilla extract

¼ cup (20 grams) powdered sugar for dusting

¼ cup chocolate frosting, Nutella, jam, preserves, Lemon Curd (page 341), or purchased dulce de leche or cajeta

EQUIPMENT

Food processor fitted with the steel blade

Baking sheets, lined with parchment paper

Fine-mesh strainer

Combine the nuts, rice and oat flours, salt, baking soda, and granulated sugar in a food processor. Pulse until the nuts are coarsely chopped. Add the cream cheese, butter, and vanilla. Process just until the dough forms a ball around the blade.

Wrap the dough and refrigerate it for at least 2 hours, but preferably overnight.

Position racks in the upper and lower thirds of the oven and preheat the oven to 325°F.

Shape slightly rounded tablespoons of dough into 1-inch balls. Place the cookies at least 1½ inches apart on the lined pans. Bake for 15 to 20 minutes, until golden brown on the bottom. Rotate the pans from top to bottom and from front to back halfway through the baking time to ensure even baking. As soon as the cookies are out of the oven, press the handle of a wooden spoon about halfway into the center of each one.

Set the pans or just the liners on racks to finish cooling. Let the cookies cool completely before storing. Unfilled cookies may be kept in an airtight container for at least 2 weeks. Use the strainer to dust the cookies with powdered sugar. Cookies may be filled in advance with frosting or Nutella, but moister fillings should be added shortly before serving to avoid making cookies soggy.

(recipe continues)

VARIATION

Walnut or Pecan Sablés

Any nut variation of this recipe makes divine sandwich cookies, filled with store-bought dulce de leche or cajeta. You can substitute an equal volume of almonds or hazelnuts for the walnuts or pecans, making almond or hazelnut sablés, if desired. Make the dough as directed but process the nuts until finely ground, rather than coarsely chopped. Form the dough into a 10- to 12-inch log 2 inches in diameter. Wrap well and chill for at least 2 hours, or overnight. Slice into scant ¼-inch-thick slices and place them 1½ inches apart on the pan. Bake for 8 to 10 minutes, until golden brown at the edges and deep brown underneath. Cool as directed. Serve plain or sandwich 2 cookies with a generous dab of dulce de leche or cajeta. The filled cookies will soften as they stand, but they are good crunchy or soft. They keep in an airtight container for at least a week.

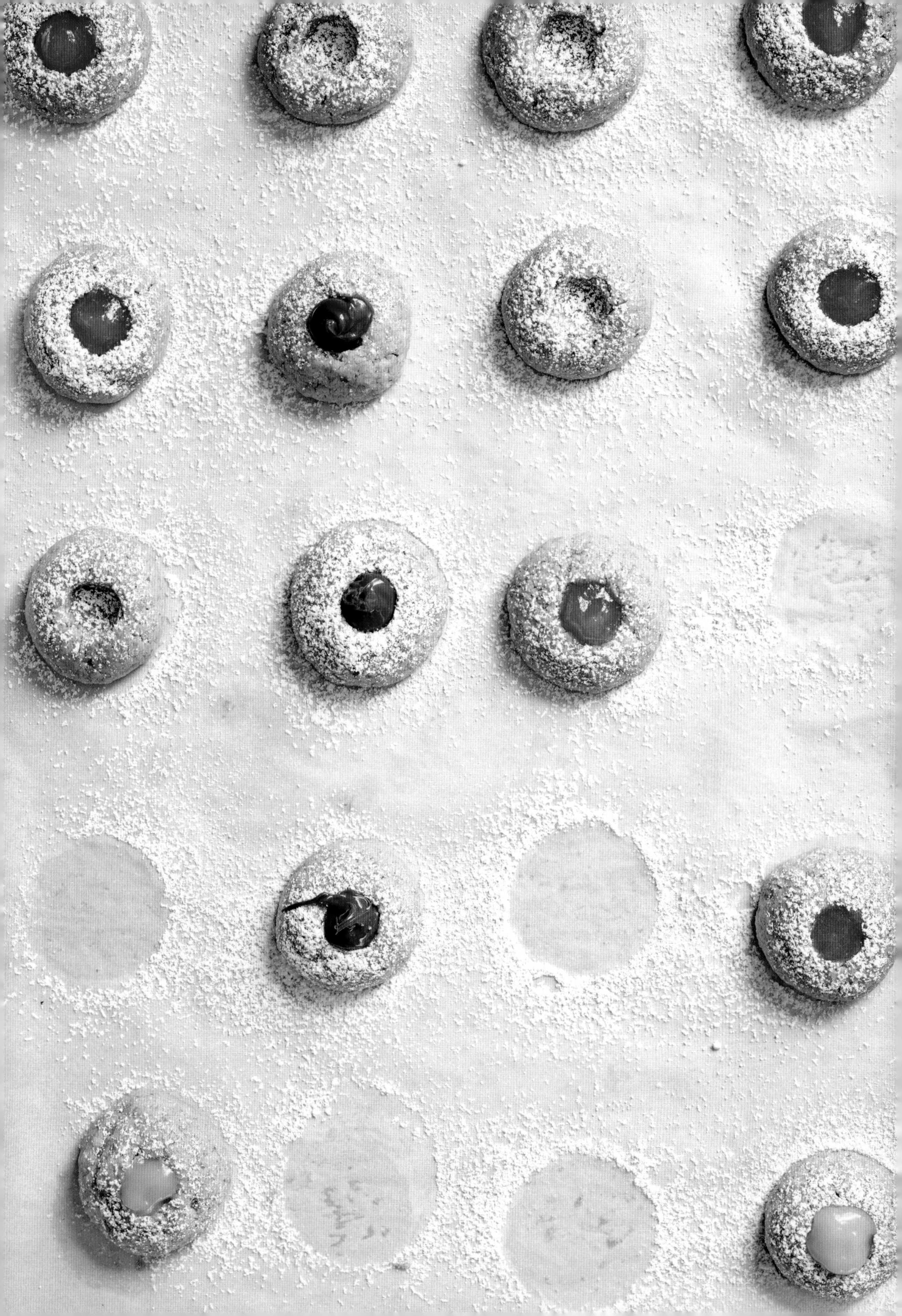

CHAPTER THREE

CORN FLOUR AND CORNMEAL

CORN is a New World grain; it has been a staple of indigenous people on the American continent for millennia and it remains a staple in our diets today. We have a long history with corn; we know it—or at least we think we do—and we certainly love it.

The best cornmeal is stone-ground and retains both the hull and the germ. Cornmeal is available in different colors and grinds. It may be ground from yellow, red, blue, or white corn (the latter more common in the American South). All of these colors of cornmeal are similar in flavor (if stone-ground) and may be used interchangeably. (Degerminated cornmeal lacks the depth of flavor and texture of stone-ground cornmeal and I do not recommend it.)

Corn flour is a very fine version of stone-ground cornmeal and thus is a whole-grain flour. It is milled from dried corn kernels to a very fine, free-flowing consistency. It feels almost as fine as wheat flour, but with a very slight grit. Pale gold or off-white (if made from white corn), corn flour has a sweet aroma reminiscent of damp straw and a sweet flavor when baked. Corn flour should not be confused with cornstarch (which is called corn flour in England and Australia). Corn*starch* is a fine white powder with no corn flavor; it is made only with the endosperm (without the germ) of the corn kernel and is used as a flavorless thickener for fruit pies and gravy and as an ingredient in many gluten-free flour blends. Cornstarch is not used in this book.

Stone-ground cornmeal and corn flour are ideal in a variety of recipes; beyond rich corn flavor and gorgeous color, both contribute an irresistible crunchy surface—or crunchy edges—to otherwise soft buttery cakes, waffles, pancakes, and even meltingly tender biscuits and cobbler toppings. Both cornmeal and corn flour are prone to grittiness unless given enough moisture and time to hydrate; recipes made with either are generally better if allowed to rest in the refrigerator, unless they contain a high percentage of moisture and cook long enough to absorb it. Because it is finer, corn flour hydrates faster and produces a less characteristically coarse "cornmeal" texture. Indeed, anyone who knows only corn bread and corn muffins will be surprised at the delicate side of corn—the fine, light-textured cakes and cookies possible from corn flour alone or in combination with a little rice flour or cornstarch. In contrast to rice and oat flours, corn has a much more pronounced flavor, pleasantly earthy and sweet—a good partner, not just a neutral background for other assertive ingredients such as berries, stone fruit, dried fruits, honey, chiles, aromatics, seeds, and spices (see sidebar).

FLAVOR AFFINITIES FOR CORN FLOUR AND CORNMEAL

Butter; dark, intense fruit, both fresh and dried; aromatics; honey; chiles; seeds

WHERE TO BUY AND HOW TO STORE

Corn flour and stone-ground cornmeal are available in better supermarkets in the specialty flour section of the baking aisle, or by mail order from Authentic Foods or Bob's Red Mill (see Resources, page 351). Both are whole grains and should be stored in an airtight container, away from heat and light, for 2 to 3 months at room temperature, or 6 months in the refrigerator, and up to 12 months in the freezer.

CORN FLOUR CHIFFON CAKE

If corn bread were transformed into a tall, moist, light and fluffy cake, this would be it. The cake is easy to make and impressive to look at. I could eat the whole cake plain, but you might want to serve slices with Honey Whipped Cream (page 345) and a little fresh grated lime zest, or with plain whipped cream and Sautéed Cherries (page 143). You can also cut the cake into three or even four layers and spread cherry jam or Lemon Curd or Lime Curd (page 341) between them. Don't miss the tart but sweet lemony variation that follows. **SERVES 10 TO 12**

---

1 cup (200 grams) sugar

5 large egg yolks, at room temperature

½ cup flavorless vegetable oil (such as corn or safflower)

½ cup water

1 cup (120 grams) corn flour

½ cup (80 grams) white rice flour
NOTE: This recipe is not successful with Thai white rice flour.

2 teaspoons baking powder

1 teaspoon salt

8 large egg whites, at room temperature

½ teaspoon cream of tartar

Lightly sweetened whipped cream (see page 343) or Honey Whipped Cream (page 345)

Sautéed Cherries (page 143)

EQUIPMENT

Stand mixer with whisk attachment

10-inch tube pan with removable bottom, ungreased

Set aside ¼ cup (50 grams) of the sugar for later (to stiffen the egg whites).

Add the remaining sugar to a large mixing bowl with the egg yolks, oil, water, corn flour, rice flour, baking powder, and salt. Whisk to blend thoroughly. Set aside for 15 minutes to hydrate the corn flour while the oven is heating.

Position a rack in the lower third of the oven and preheat the oven to 325°F.

Combine the egg whites and cream of tartar in the bowl of the stand mixer and beat with the whisk attachment on medium-high speed until the egg whites are creamy white and hold a soft shape when the beaters are lifted. Slowly sprinkle in the reserved sugar, beating at high speed until the egg whites are stiff but not dry. Scrape half of the egg whites onto the batter and fold until partially blended. Add the remaining egg whites and fold just until the batter looks blended. Scrape the batter into the pan and spread it evenly.

Bake for 45 to 50 minutes, until the top of the cake is golden brown and a toothpick inserted in the center comes out clean.

Set the pan on a rack. While the cake is still hot, slide a thin knife or spatula around the sides, pressing against the pan to avoid tearing the cake. Leave the cake in the pan to cool—it will settle at least 1 inch.

When cool, lift the tube to remove the cake. Slide a thin knife or skewer around the tube and slide a spatula under the cake all around. Lift the cake off the bottom of the pan using two spatulas (one on either side of the tube) and transfer it to a serving platter.

The cake keeps, wrapped airtight, at room temperature for at least 3 days, or in the freezer for up to 3 months; bring to room temperature before serving.

Slice with a serrated knife. Serve with whipped cream and any of the accompaniments mentioned above.

VARIATION: Lemon and Corn Flour Chiffon Cake

Grate the zest of one large unsprayed or organic lemon into the bowl with the other ingredients and proceed as directed. While the cake is in the oven, mix 6 tablespoons of fresh lemon juice with 1¼ cups (140 grams) powdered sugar. When the cake is done, poke the top all over with a toothpick. Spoon the lemon mixture over the hot cake. Cool in the pan as directed.

GOLDEN CORN CAKE

The top of this buttery rich cake settles down slightly to resemble a velvety golden brown beret. The color is all corn bread, but the texture is softer and moister than that of pound cake. Serve slices (with or without a splash of rum or tequila) with whipped cream and fresh lime zest grated over the top. Or opt for whipped cream and Sautéed Cherries (opposite). **SERVES 8 TO 10**

---

8 tablespoons (115 grams) Clarified Butter (page 335) or ghee (see page 38)

⅔ cup (80 grams) corn flour

2 tablespoons (20 grams) white rice flour
NOTE: This recipe is not successful with Thai white rice flour.

2 large eggs

6 large egg yolks

⅔ cup (130 grams) granulated sugar

¼ teaspoon salt

1 teaspoon pure vanilla extract

Powdered sugar for dusting (optional)

Sautéed Cherries (recipe follows; optional)

Lightly sweetened whipped cream (see page 343; optional)

EQUIPMENT

8-by-3-inch springform pan or cheesecake pan with removable bottom, ungreased

Stand mixer with whisk attachment

Sifter or medium-mesh strainer

Position a rack in the lower third of the oven and preheat the oven to 350°F.

Put the clarified butter in a small pot or microwavable container ready to reheat when needed, and have a 4- to 5-cup bowl ready to pour it into as well—the bowl must be big enough to allow you to fold some batter into the butter later.

Whisk the corn flour and rice flour together thoroughly in a medium bowl.

Combine the eggs, egg yolks, granulated sugar, salt, and vanilla in the bowl of the stand mixer and beat with the whisk attachment on high speed for at least 5 minutes. The mixture should be light colored and tripled in volume, and you should see well-defined tracks as the whisk spins; when the whisk is lifted, the mixture should fall in a thick, fluffy rope that holds its shape for several seconds on the surface of the batter.

Just before the eggs are ready, heat the clarified butter until very hot and pour it into the reserved bowl.

Remove the bowl from the mixer. Sift one-third of the flour over the eggs. Fold with a large rubber spatula until the flour is almost blended into the batter. Repeat with half of the remaining flour. Fold in the rest of the flour. Scrape about one-third of the batter into the hot butter. Fold until the butter is completely blended into the batter. Scrape the buttery batter over the remaining batter and fold just until blended. Scrape the batter into the pan and tilt to level if necessary.

Bake for 35 to 40 minutes, or until golden brown on top and a toothpick inserted in the center comes out clean and dry. Set the pan on a rack. While the cake is still hot, run a small spatula around the inside of the pan, pressing against the sides of the pan to avoid tearing the cake. Let cool completely. Remove the pan sides and transfer the cake to a serving platter.

The cake may be wrapped airtight and stored at room temperature for 2 to 3 days, or frozen for up to 3 months. Dust with powdered sugar, if desired, and serve plain or with whipped cream and sautéed cherries.

Sautéed Cherries

MAKES ABOUT 2 CUPS

1 pound (3 cups/455 grams) fresh cherries

3 tablespoons red wine or water

1 tablespoon sugar

1 teaspoon lemon juice

Pinch of salt

Pit and halve the cherries; you should have about 2½ cups. Toss them in a bowl with the wine, sugar, lemon juice, and salt. Let stand for 5 minutes or so to dissolve the sugar. Heat a wide nonreactive skillet over medium-high heat until a cherry sizzles when you toss it in. Add the cherries and their juices and sauté them, scraping the pan frequently to prevent the juices from burning, just long enough to reduce the juices to a thick, sticky glaze. This should take only a couple of minutes; the cherries should be slightly cooked, not mushy. (If the pan is too crowded or the heat too low, the cherries will simmer too long in too much juice.) Scrape the cherries and syrup into a bowl and refrigerate or set aside until completely cool before using. The cherries may be kept in an airtight container in the refrigerator for at least 1 week.

NEW CLASSIC BOSTON CREAM PIE

Does everyone know that a Boston cream pie is a cake and not a pie at all? Normally it has two layers of white or yellow cake with a luxurious amount of vanilla custard (aka pastry cream) between them, and a thick chocolate glaze. Think a chocolate-glazed vanilla custard–filled éclair, but with cake instead of cream puff pastry. This new "pie" is still a cake, but a lighter and more flavorful corn flour chiffon cake. **SERVES 10 TO 12**

½ cup (100 grams) sugar

3 large egg yolks, at room temperature

¼ cup cool water

¼ cup flavorless vegetable oil (such as corn or safflower)

⅓ cup (40 grams) corn flour

⅓ cup (50 grams) white rice flour
NOTE: This recipe is not successful with Thai white rice flour.

1 teaspoon baking powder

Scant ½ teaspoon salt

4 large egg whites, at room temperature

¼ teaspoon cream of tartar

2 cups The New Vanilla Pastry Cream (page 342)

Cocoa Fudge Glaze (recipe follows)

EQUIPMENT

Two 8-inch round cake pans

Stand mixer with whisk attachment

Baking sheet

Set aside 2 tablespoons (25 grams) of the sugar for later (to stiffen the egg whites).

In a large bowl, whisk the remaining sugar with the egg yolks, water, oil, corn and rice flours, baking powder, and salt until thoroughly blended. Set aside for 15 minutes to hydrate the flour while the oven is heating.

Position a rack in the lower third of the oven and preheat the oven to 325°F. Line the bottoms of the pans with parchment paper.

In the bowl of the stand mixer, beat the egg whites and cream of tartar with the whisk attachment at medium-high speed until the egg whites are creamy white and hold a soft shape when the beaters are lifted. Slowly sprinkle in the reserved sugar, beating at high speed until the egg whites are stiff but not dry. Scrape one-quarter of the egg whites onto the batter and use a rubber spatula to fold them in. Fold in the remaining egg whites. Divide the batter evenly between the pans and spread it evenly.

Bake for 20 to 25 minutes, until the tops of the cakes are golden brown and spring back when you press gently with a finger, and a toothpick inserted in the center comes out clean. Rotate the pans in the oven about halfway through the baking. Set the pans on a rack to cool.

At your convenience (the cake can be warm or completely cool), run a small spatula around the inside of the pans, pressing against the sides of the pan to avoid tearing the cake. Invert the pans onto the rack to remove the cakes and peel off the parchment liner. Turn the cakes right side up to finish cooling. The cakes may be wrapped airtight and stored at room temperature for 2 to 3 days, or frozen for up to 3 months.

(recipe continues)

Set one cake layer on a cardboard cake circle or the bottom of a springform pan (so it will be easier to lift onto the platter after glazing). Spread all of the pastry cream evenly over the layer. Top with the second cake layer, bottom side up. If not glazing and serving within the hour, cover and refrigerate for up to 1 day.

If the glaze has cooled too much to pour, warm it gently in a pan of barely simmering water until it is fluid but not too thin to coat the cake luxuriously. Set the cake (still on the cake circle or pan) on the baking sheet. Pour the glaze over the cake, leaving some of the sides exposed if you like. Transfer the cake to a serving platter and refrigerate. Remove the cake from the refrigerator 60 minutes before serving.

Cocoa Fudge Glaze

If you use this while it is warm, it makes a luscious, thick glaze (or sauce for ice cream). You can also cool it and use it as a frosting. **MAKES 2 CUPS**

1 cup (200 grams) sugar

1 cup (90 grams) unsweetened cocoa powder

A pinch or two of salt

1 cup heavy cream

6 tablespoons (85 grams) unsalted butter

2 teaspoons pure vanilla extract

Put the sugar, cocoa, and salt in a medium saucepan and stir in just enough of the cream to make a smooth, thick paste, then stir in the rest of the cream. Add the butter and stir over low heat until the butter is melted and the mixture is smooth and very hot but not simmering. Taste and stir in a little more salt if desired. Remove from the heat and stir in the vanilla.

Let cool until lukewarm but still pourable. Store leftover glaze in the refrigerator; rewarm gently in a pan of barely simmering water or in a microwave before using. Glaze may be refrigerated, covered, for a week or frozen for up to 3 months.

CORN FLOUR BISCUITS

With an appealing crunchy crust, these biscuits are soft inside and have a gentler corn flavor than corn bread. Serve them warm with a special honey or berry jam, or with dry-cured ham or a spicy cheese spread. Biscuits have the best texture and rise better if the dough rests in the refrigerator for at least 2 hours before baking, either as a whole log or sliced. **MAKES TWELVE 3-INCH BISCUITS**

1⅓ cups (200 grams) white rice flour (preferably superfine)
—OR—
2 cups (200 grams) Thai white rice flour

½ cup (60 grams) corn flour

2 teaspoons sugar

¼ teaspoon xanthan gum

1 tablespoon baking powder

½ teaspoon salt

1 cup heavy cream

½ cup plain yogurt (any percent fat) or slightly watered down Greek yogurt

EQUIPMENT

Stand mixer with paddle attachment

Baking sheet, lined with parchment paper

Combine the rice flour, corn flour, sugar, xanthan gum, baking powder, salt, cream, and yogurt in the bowl of the stand mixer and beat with the paddle attachment for 2 minutes on low speed; the dough will be very stiff. It is important to beat the dough long enough or the biscuits won't rise well; don't worry about overbeating.

Form the dough into a log 2 inches in diameter, wrap it in plastic, and refrigerate for at least 2 hours or up to 2 days.

Position a rack in the upper third of the oven and preheat the oven to 400°F.

Cut the log into 12 thick slices. Place the slices close together on the pan for soft pull-apart biscuits or 2 inches apart for separate biscuits and bake for 20 to 25 minutes until browned on both top and bottom. Serve immediately or cool on a rack and toast before serving.

CORN FLOUR TEA CAKE WITH CURRANTS AND PISTACHIOS

This not-too-sweet quick bread, or teatime loaf cake, is crusty and crunchy on the outside and moist and buttery within. The festive golden cake is laced with pistachios, currants, and fragrant aniseed. After a day or so, refresh slices for a few seconds in the microwave, or butter and then toast them—so decadent! Or toast slices and top with just a drizzle of honey or smear of marmalade. Serve with coffee or black tea with milk. **SERVES 6 TO 8**

---

1 cup plus 3 tablespoons (180 grams) white rice flour
—OR—
1¾ cups (180 grams) Thai white rice flour

¼ cup plus 2 tablespoons (50 grams) corn flour

¾ cup plus 2 tablespoons (180 grams) sugar

8 tablespoons (1 stick/115 grams) unsalted butter, very soft

Scant ½ teaspoon salt

1 teaspoon baking powder

½ teaspoon baking soda

¼ teaspoon xanthan gum

¾ cup plain yogurt (any percent fat) or slightly watered down Greek yogurt

2 large eggs

1 teaspoon pure vanilla extract

½ teaspoon aniseed

½ cup dried currants

½ cup (50 grams) chopped toasted pistachio nuts

EQUIPMENT

Stand mixer with paddle attachment

8½-by-4½-inch (6-cup) loaf pan, bottom and all four sides lined with parchment paper

Position a rack in the lower third of the oven and preheat the oven to 350°F.

Combine the rice and corn flours, sugar, butter, and salt in the bowl of the stand mixer and beat on medium speed with the paddle attachment until the mixture has the texture of brown sugar, about a minute. Add the baking powder, baking soda, xanthan gum, yogurt, eggs, vanilla, and aniseed and beat at medium-high speed for 2 to 3 minutes; the batter should be very smooth and fluffy. Beat in the currants and nuts on low speed.

Scrape the batter into the prepared pan and bake for 50 to 55 minutes, or until a toothpick inserted in the center comes out clean and dry. Cool in the pan on a rack. The cake keeps for up to 3 days at room temperature in an airtight container.

CORN FLOUR AND CRANBERRY SCONES

These gently sweet scones with their all-American flavors don't really *need* a holiday, but they are excellent stand-ins for plain dinner rolls on the Thanksgiving table. The dough can be made ahead and ready to bake while your bird is resting. Or bake and serve the scones with leftover turkey later in the week. Scones will have the best texture and rise if the dough rests in the refrigerator for at least 2 hours before baking, either as a whole log or sliced. **MAKES TWELVE 3-INCH SCONES**

1⅓ cups (200 grams) white rice flour (preferably superfine)
—OR—
2 cups (200 grams) Thai white rice flour

¼ cup plus 2 tablespoons (60 grams) stone-ground cornmeal

¼ cup (50 grams) granulated sugar

¼ teaspoon xanthan gum

1 tablespoon baking powder

½ teaspoon salt

1 cup heavy cream

½ cup plain yogurt (any percent fat) or slightly watered down Greek yogurt

1 cup (120 grams) dried cranberries

About 2 tablespoons coarse sugar, such as turbinado, for the tops of the scones

EQUIPMENT

Stand mixer with paddle attachment

Large baking sheet, lined with parchment paper

Combine the rice flour, cornmeal, granulated sugar, xanthan gum, baking powder, salt, cream, and yogurt in the bowl of the stand mixer and beat with the paddle attachment for 2 minutes on low speed, or until thoroughly mixed and smooth. It is important to beat the dough enough or the scones won't rise well; don't worry about overbeating. Beat in the cranberries.

Form the dough into a log 2 inches in diameter, wrap it in plastic, and refrigerate for at least 2 hours or up to 2 days.

Position a rack in the upper third of the oven and preheat the oven to 400°F.

Cut the log into 12 thick slices. Place the slices 2 inches apart on the lined pan and sprinkle with the coarse sugar. Bake for 20 to 25 minutes, or until the scones are browned on both top and bottom. Serve immediately or cool on a rack and toast before serving.

VARIATION: Corn and Blueberry Scones

Substitute 1 cup (120 grams) dried blueberries for the cranberries.

CRUNCHY CORN FRITTERS

Corn fritters go from morning to evening, sweet or savory. Serve them with honey or a drizzle of sorghum syrup for breakfast or brunch, or offer them plain for lunch with a big summer salad. At cocktail time, pass them with a bowl of Sriracha "aïoli": mayonnaise flavored to your taste with Sriracha (aka Rooster Sauce). **SERVES 6 TO 8 AS AN APPETIZER**

---

1 teaspoon sugar

¼ cup plus 2 tablespoons water

2 tablespoons (30 grams) unsalted butter

½ teaspoon salt

⅓ cup (40 grams) corn flour

1 large egg

1 teaspoon baking powder

1 tablespoon bourbon or water

¼ cup (40 grams) glutinous rice flour
—OR—
⅓ cup (40 grams) Thai glutinous white rice flour

1 cup (200 grams) drained cooked corn kernels (fresh or frozen and defrosted)

1 to 1½ quarts vegetable oil, such as corn or peanut

Coarse salt for sprinkling (optional)

Sriracha aïoli (see headnote)

EQUIPMENT

Stand mixer with paddle attachment

Deep-fat fryer or medium (2- to 3-quart) saucepan and frying thermometer

Combine the sugar, water, butter, and salt in a small saucepan and bring to a boil over medium heat. Add the corn flour and stir with a long-handled metal or wooden spoon until smooth. Turn the heat to low and push the dough around the pan for 2 more minutes, turning the dough over in the pan to avoid scorching. Scrape the dough into the mixer bowl. Break the egg into the still-hot saucepan and swirl to warm the egg.

Turn the mixer to medium speed and add the egg, beating until the dough is glossy and smooth, about 2 minutes. Scrape down the sides of the bowl and add the baking powder, bourbon, and glutinous rice flour. Mix on medium speed until the dough is very smooth and elastic, another minute or so. Stir in the corn.

Pour oil to a depth of about 2 inches in the deep-fat fryer or saucepan and heat to 350°F. Using two spoons or a small spring-loaded scoop, place 1½-teaspoon-sized lumps of batter in the oil. Do not crowd the pan or fryer; the dough will expand about eightfold. After a minute or so, use long-handled tongs to turn the fritters. Fry until very brown on all sides, 3 to 5 minutes. If necessary to test doneness, cut a fritter in half. Drain the fritters on a cake rack; repeat with the remaining batter.

Serve immediately, sprinkled with coarse sea salt, if desired, and with Sriracha aïoli on the side. If the fritters have cooled off, reheat for 5 minutes in a 400°F oven before serving. Leftovers may be stored, covered loosely with a paper towel, at room temperature for up to 2 days before reheating.

LEMON TART

Corn flour adds great flavor and a little extra crunch to everyone's favorite citrus tart. Leftover lemon curd filling is a bonus: It keeps for several days and is delicious on toast or scones, or spread on a butter cookie. SERVES 8 TO 10

FOR THE CRUST

½ cup plus 1 tablespoon (65 grams) corn flour, preferably superfine

¼ cup (40 grams) white rice flour
—OR—
⅓ cup plus 1 tablespoon (40 grams) Thai white rice flour

¼ cup (50 grams) sugar

⅛ teaspoon salt

1/16 teaspoon baking soda

6 tablespoons (85 grams) unsalted butter, slightly softened and cut into chunks

2 tablespoons (30 grams) cream cheese

2 teaspoons water

½ teaspoon pure vanilla extract

FOR THE FILLING

4 large eggs

3 large egg yolks

Grated zest of 2 medium lemons

1 cup strained fresh lemon juice

1 cup (200 grams) sugar

12 tablespoons (1½ sticks/170 grams) unsalted butter

EQUIPMENT

Food processor fitted with the steel blade (optional)

9½-inch fluted tart pan with a removable bottom

Rimmed baking sheet

Medium-mesh strainer

Grease the tart pan with vegetable spray or butter. To make the crust by hand, put the corn and rice flours, sugar, salt, and baking soda in a medium bowl and whisk until thoroughly blended. Add the butter, cream cheese, water, and vanilla. Use a fork or the back of a large spoon to mash and mix the ingredients together until all are blended into a smooth, soft dough.

To make the crust in a food processor, put the corn and rice flours, sugar, salt, and baking soda in the food processor. Pulse to blend. Add the butter, cream cheese, water, and vanilla. Pulse until the mixture forms a smooth, soft dough. Scrape the bowl and blend in any stray flour at the bottom of the bowl with your fingers.

Transfer the dough to the tart pan.

The dough may seem much softer than other tart doughs. Use the heel of your hand and then your fingers and/or a small offset spatula to spread the dough all over the bottom of the pan. Press it squarely into the corners of the pan with the side of your index finger to prevent extra thickness at the bottom edges, and press it as evenly as possible up the sides, squaring it off along the top edge. Have patience; there is just enough dough (although you may not think so at first). If there is too much dough in one place (or hiding in the corners of the pan), pinch or scrape it off and move it elsewhere. Spread or smear it smooth with the spatula. Here's a final trick for a perfectly even crust: Press a sheet of plastic wrap against the bottom and up the sides of the pan and lay a paper towel on top. Set a straight-sided, flat-bottomed cup on the towel; press and slide the cup all over the bottom and into the corners to smooth and even the surface. Leave the plastic wrap in place. Refrigerate the pan for at least 2 hours, but preferably overnight and up to 3 days.

NOTES: There is plenty of time to make the lemon filling while the crust is baking, but should there be a delay, don't worry. Just take the crust out of the oven when it is done and set it aside, leaving the oven on. When the filling is ready, proceed as directed.

You can make the lemon curd filling in advance, too. It keeps in the refrigerator for several days. To use it in the tart, simply add about 5 minutes to the final baking time to reheat the chilled filling.

Position a rack in the lower third of the oven and preheat the oven to 350°F.

Peel off the plastic wrap and place the pan on the baking sheet. Bake for 30 to 35 minutes, checking after 15 or 20 minutes. If the crust has puffed up on the bottom, press it back down carefully with the back of a fork. Continue baking until the crust is golden brown and has slightly pulled away from the edges of the pan all around.

Meanwhile, to make the filling, set the strainer over a medium bowl near the stove. Whisk the eggs and yolks in a medium nonreactive saucepan to blend. Whisk in the lemon zest, juice, and sugar. Add the butter. Whisk over medium heat, reaching into the corners and scraping the sides and bottom of the pan, until the butter is melted and the mixture is thickened and beginning to simmer around the edges; then continue to whisk for about 10 seconds longer. Remove from the heat and scrape the filling into the strainer, pressing gently on the solids. Scrape any lemon curd clinging to the underside of the strainer into the bowl. Set aside until needed.

When the crust is ready, remove it from the oven, leaving the oven on and the door closed. Spread the lemon filling evenly over the crust (you may have some left over) and return the tart to the oven for 5 minutes. (Leftover lemon curd keeps, covered airtight, in the refrigerator for several days.) Set the pan on a rack to cool.

Serve the tart on the day you make it, at room temperature or chilled. To serve, remove the sides of the pan and transfer to a platter. Leftovers keep, covered, in the refrigerator for a day or so, although the crust may soften.

THE NEW CHOCOLATE CREAM PIE

Thick, creamy, and smooth dark chocolate pudding in a crunchy cinnamon and cornflake crust makes for a crowd-pleasing dessert. Top it with whipped cream and a little freshly grated cinnamon stick (and maybe a handful of toasted pecans); it's gorgeous as well. SERVES 8 TO 10

FOR THE CRUST

3 cups (110 grams) cornflakes

3 tablespoons (35 grams) sugar

⅛ teaspoon salt

½ teaspoon ground cinnamon

4 tablespoons (½ stick/55 grams) unsalted butter, melted

3 ounces (85 grams) milk chocolate, finely chopped

FOR THE FILLING

1 recipe Silky Chocolate Pudding (page 80)

Lightly sweetened whipped cream (see page 343)

Cinnamon stick, for grating (optional)

EQUIPMENT

Food processor fitted with the steel blade

9-inch glass or ceramic pie plate

Position a rack in the lower third of the oven and preheat the oven to 350° F. To make the crust, pulverize the cornflakes to fine crumbs in the food processor. Add the sugar, salt, cinnamon, and butter and process until well blended.

Scrape the mixture into the pie plate and pat it evenly across the bottom, up the sides, and partially onto the rim. Press firmly with the bottom and sides of a small cup to compact the mixture and adhere it to the pan. Bake for 10 to 15 minutes, pressing gently with the back of a fork or the cup if the crust puffs up, until the crust looks set and a shade darker than the unbaked crumbs and the edges are browned. Remove the crust from the oven and sprinkle the milk chocolate over the bottom. Wait 3 to 4 minutes for the chocolate to soften. Use the back of a spoon to spread the chocolate over the bottom and sides of the crust. Let cool and refrigerate briefly to set the chocolate if necessary.

Make the filling as directed.

Scrape the pudding into the piecrust (pouring any extra into a small bowl). Let cool for 1 hour, then refrigerate the pie, uncovered, for at least an hour to be sure the filling is completely cool. If not serving shortly, cover the pie and return it to the refrigerator.

Top the pie with whipped cream and sprinkle with grated cinnamon, if you like. Leftovers may be refrigerated, covered, for 2 or 3 days.

BLUEBERRY CORN FLOUR COBBLER

Tender yet crunchy corn biscuits make the best and tastiest topper for blueberry cobbler—or one made with any dark berry or stone fruit. These assertive fruit flavors play brilliantly with the sweet and earthy corn. This cobbler is distinctive and superb just as it is, but you can always dress it up with a scoop of vanilla ice cream or whipped cream. **SERVES 6 TO 8**

FOR THE FRUIT

5 cups (700 grams) fresh or frozen blueberries (not defrosted)

2 tablespoons lemon juice

1 tablespoon white rice flour

½ cup (100 grams) sugar

FOR THE BISCUIT TOPPING

⅔ cup (100 grams) white rice flour (preferably superfine)
—OR—
1 cup (100 grams) Thai white rice flour

¼ cup (30 grams) corn flour

⅛ teaspoon xanthan gum

1 tablespoon granulated sugar

1½ teaspoons baking powder

¼ teaspoon salt

½ cup heavy cream

¼ cup plain yogurt (any percent fat) or slightly watered down Greek yogurt

1 tablespoon coarse sugar, such as turbinado, for sprinkling

EQUIPMENT

2-quart baking dish, 2 to 3 inches deep

Rimmed baking sheet

Stand mixer with paddle attachment

Position a rack in the lower third of the oven and preheat the oven to 400°F.

For the fruit, combine the berries, lemon juice, rice flour, and sugar in the baking dish. Place the dish on a rimmed baking sheet and bake for 15 minutes. Stir, then bake for 15 to 20 more minutes, or until juices are bubbling at the edge of the dish.

Mix the biscuit dough while the fruit is baking: Combine the rice and corn flours, xanthan gum, granulated sugar, baking powder, salt, cream, and yogurt in the bowl of the stand mixer and beat with the paddle attachment for 2 minutes on low speed; the dough will be very stiff. It is important to beat the dough enough or the biscuits won't rise well; don't worry about overbeating.

When the berries are ready, spoon dollops of biscuit dough over them—don't cover the fruit completely—and sprinkle with the coarse sugar. Bake for 15 to 20 minutes, or until the top is browned and the filling is bubbling in the center. Cool for about 20 minutes and serve warm or at room temperature. The cobbler keeps at room temperature for up to 1 day, covered with a paper towel, or refrigerated for up to 3 days, covered with plastic wrap.

VARIATIONS

Blackberry Corn Flour Cobbler

Substitute 5 cups (700 grams) fresh or frozen (not defrosted) blackberries for the blueberries.

Peach or Prune Plum Cobbler

Substitute 5 cups (775 grams) peach slices (from about 2 pounds/900 grams peaches) or quartered prune plums for the blueberries.

SWEET OR SAVORY CORN STICKS

These light, dry, super-crunchy corn cookies are subtly sweet to start with. Drizzle them with a few drops of extra-virgin olive oil and sprinkle them with flaky salt—with or without sweet, hot, or smoky paprika—and you've got an addictive savory snack or cocktail bite. **MAKES 24 STICKS**

---

3 tablespoons (45 grams) unsalted butter

¾ cup plus 2 tablespoons (110 grams) corn flour

⅜ cup (75 grams) sugar

Generous ⅛ teaspoon salt

½ teaspoon baking powder

2 large eggs, at room temperature

EQUIPMENT

Heavy skillet at least 12 inches wide

8-inch square baking pan, bottom lined with parchment paper

Stand mixer with whisk attachment, or handheld mixer

Baking sheet

Melt the butter in a large heavy-bottomed skillet. Take it off the heat, add the corn flour, and stir to coat all of the flour grains with butter. The mixture will have the consistency of slightly damp sand. Return the skillet to the stove and cook over medium-high heat, stirring constantly with a heatproof spatula or fork; scrape the bottom and sides of the pan, turning the flour, and spread or rake to redistribute it continuously so that it toasts evenly. Continue to cook and stir until the mixture colors slightly and smells toasted; it may begin to smoke a little. Toasting the flour will take 4 to 6 minutes. Scrape the flour into the lined baking pan and spread it out to cool while preheating the oven.

Position a rack in the lower third of the oven and preheat the oven to 350°F.

Combine the sugar, salt, baking powder, and eggs in the bowl of the stand mixer fitted with the whisk attachment (or in a large bowl if using a handheld mixer). Beat on high speed for 3 to 5 minutes, until thick and light. Poke and mash any large lumps in the toasted flour and then pour it over the egg mixture. Fold just until evenly mixed. Scrape the batter into the lined baking pan and spread it evenly; it should be a thin layer only about ½ inch deep.

Bake for 15 to 20 minutes, or until golden brown and springy to the touch. Cool on a rack. Lower the oven temperature to 300°F. Slide a slim knife around the edges of the pan to detach the baked sheet. Invert the pan onto a rack and peel off the liner, then turn the sheet right side up on a cutting board. Cut it in half lengthwise with a sharp serrated knife. Cut each half crosswise in slices a scant ¾ inch wide. Arrange the slices slightly apart, standing up, on an unlined baking sheet. Bake at 300°F for 20 to 25 minutes, or until slightly golden brown. Rotate the sheet from front to back about halfway through the baking time.

(recipe continues)

Cool the sticks completely before storing. They may be stored in an airtight container for several weeks.

VARIATIONS

Brown Butter Toasted Corn Sticks with Extra-Virgin Olive Oil and Sea Salt

After baking, drizzle the corn sticks with extra-virgin olive oil and sprinkle with pinches of flaky sea salt (if the flakes are very large, crush the pinches slightly before sprinkling).

Brown Butter Toasted Corn Sticks with Smoky Paprika

After baking, drizzle the corn sticks with extra-virgin olive oil, as directed above, followed by pinches of smoked, sweet, or hot paprika.

SEED CRACKERS

These superthin, dramatically big, and addictively delicious crackers are loaded with good grains and seeds. Break them into rustic shards or leave them whole, but definitely serve with savory spreads and cheeses. **MAKES 1½ DOZEN LARGE CRACKERS**

½ cup plus 1 tablespoon (80 grams) brown rice flour

½ cup (80 grams) white rice flour
—OR—
¾ cup (80 grams) Thai white rice flour

¼ cup plus 2 tablespoons (40 grams) gluten-free oat flour

⅔ cup (80 grams) stone-ground cornmeal

½ cup (65 grams) roasted unsalted sunflower seeds

¼ cup (35 grams) sesame seeds, toasted (see page 309)

¼ cup plus 2 tablespoons (40 grams) flaxseed meal, or ¼ cup (40 grams) whole flaxseeds, finely ground

2 tablespoons (25 grams) packed light or dark brown sugar

1½ teaspoons salt

¾ cup plus 1 tablespoon water

2 teaspoons rice vinegar

2 teaspoons baking powder

¼ cup corn oil

Flaky salt or other coarse salt for sprinkling (about 1 tablespoon)

EQUIPMENT

Stand mixer with paddle attachment

Baking sheets

Rolling pin

Position racks in the upper and lower thirds of the oven and preheat the oven to 450°F.

Mix the brown rice, white rice, and oat flours, cornmeal, sunflower and sesame seeds, flaxseed meal, brown sugar, and table salt in the bowl of the stand mixer fitted with the paddle attachment. Add the water and vinegar and beat for 2 minutes on medium speed to form a thick, sticky dough that might wrap around the paddle at first. Sprinkle in the baking powder and add the corn oil; beat for 1 minute on medium speed to thoroughly incorporate the oil.

Cut four pieces of parchment the size of a baking sheet. Drop three 2-tablespoon lumps of dough evenly spaced down the length of a parchment sheet. Cover with another piece of parchment and flatten each lump with the heel of your hand. Use a rolling pin to roll the dough into oblongs from about 3 by 5 inches to 4 by 8 inches (for even thinner crackers) and a scant ⅛ inch thick. Peel off the top parchment (save for reuse), sprinkle the dough very lightly with flaky salt, and place the parchment with the crackers *dough side down* on a baking sheet.

Bake for 5 to 6 minutes, two pans at a time, rotating them from upper to lower and front to back for even baking, or until the crackers are browned at the edges. Remove the pan from the oven and carefully peel off the parchment (save for reuse), using a spatula to ease the paper off, if necessary. Flip the crackers over with a spatula and return them to the oven for 2 to 3 minutes, or until well browned at the edges. Repeat with the remaining dough: while the crackers are baking, continue to roll out more dough; as soon as the pans of crackers are done, flip the next batch onto the baking sheets (it's okay if the pans are still hot as long as you put them in the oven immediately).

Cool the crackers thoroughly on a rack before storing in an airtight container for up to 10 days. The crackers may be refreshed (if they are not crispy) before serving by baking for 5 minutes at 400°F.

SOUFFLÉED CORN FLOUR AND YOGURT PUDDINGS WITH CAJETA

Cajeta is goat's milk caramel (found in specialty food shops or Latin American groceries). I love to pair this yummy gutsy Mexican sauce with a simple baked pudding made with goat's-milk yogurt. But you can dial down the exotic and aim for crowd-pleasing instead: just substitute dulce de leche for cajeta and regular yogurt for the goat's-milk version. These are simple to make but must be prepared just before serving. Don't miss the more chocolaty variation. SERVES 6

Softened or melted unsalted butter and sugar for coating the bowls

One 8-ounce container full-fat goat's-milk yogurt

3 large eggs, separated, at room temperature

3 tablespoons (25 grams) corn flour

⅛ teaspoon salt

⅛ teaspoon cream of tartar

5 tablespoons (60 grams) sugar, plus more for sprinkling

1 ounce (30 grams) milk chocolate, chopped

Cinnamon stick, for grating (or add a pinch of ground cinnamon to the sugar for sprinkling)

⅔ cup cajeta or dulce de leche

EQUIPMENT

Six 8- to 10-ounce ovenproof bowls

Rimmed baking sheet

Stand mixer with whisk attachment, or handheld mixer

Position a rack in the center of the oven and preheat the oven to 375°F. Butter and sugar the bowls and set them on the baking sheet.

Combine the yogurt, egg yolks, flour, and salt in a large bowl and stir until blended.

Combine the egg whites and cream of tartar in the bowl of the stand mixer (or in another large bowl if using a handheld mixer) and beat on medium speed (or on high speed with the handheld mixer) until the egg whites are creamy white and hold a soft shape when the beaters are lifted. Gradually add the 5 tablespoons sugar, continuing to beat on high speed until the egg whites are stiff but not dry. Fold about one-quarter of the egg whites into the yogurt mixture, then fold in the remaining egg whites.

Divide the batter among the bowls. Sprinkle the top of each bowl with a pinch or two of sugar and some of the chopped chocolate. Grate a little cinnamon on top. Bake for 15 to 20 minutes, until puffed and slightly golden brown. While the puddings are baking, warm the cajeta just until the sauce is pourable. Let the puddings sit for about 5 minutes, then drizzle with a little of the sauce before serving. Pass a pitcher or bowl of the remaining cajeta and let guests help themselves.

VARIATION: Souffléed Corn Flour Puddings Laced with Milk Chocolate

Fold ⅓ cup (55 grams) finely chopped milk chocolate into the batter with the final addition of egg whites.

CHAPTER FOUR

BUCKWHEAT FLOUR

BUCKWHEAT is not a type of wheat or even a grass; therefore it is not a true grain. Sometimes referred to as a pseudo-cereal—since its seeds are cooked like cereal and made into flour as well—buckwheat is actually related to sorrel and rhubarb.

Buckwheat is most familiar in Japanese soba (noodles), French crepes from Brittany, and hearty buckwheat pancakes, including Russian blini, as well as the beloved (or detested) Jewish grain dish called kasha. Buckwheat's reputation for robust earthy flavors—an acquired taste for some—surely comes from these iconic dishes. But different approaches and different recipes coax different characteristics from buckwheat. Adventuresome pastry chefs have expanded the use of buckwheat in the last decade to include some more sophisticated and delicate cakes and cookies with a less ethnic appeal than soba or your bubbe's kasha.

Buckwheat flour is made by first roasting and then milling the starchy, pyramid-shaped buckwheat seeds to a fine, slightly crystalline powder. Slate-y lavender brown and speckled with bits of black hull, buckwheat flour (before baking) has a toasty aroma with pleasing but slightly sour or fermented vegetal notes of green wood, oak, grass, and green tea. Because buckwheat is a very assertive flavor, it is usually blended with wheat flour; I mix it with rice flour instead or use it alone in recipes that call for only small amounts of flour.

Buckwheat provides a perfect example of how any given flour can produce completely contradictory results, depending on the type of recipe it is used in and how it is handled. Experimenting with buckwheat flour in light sponge batters and génoise, I was consistently rewarded with delightful floral flavors and light, slightly crystalline textures. But experimenting with a fine-grained butter cake in her kitchen 600 miles away, Maya Klein was getting earthy flavors and dense gummy textures from the very same flour. Which of us was hallucinating? Maya thought the ultrafine texture of the buckwheat flour was the culprit; to prove it, she ground some roasted buckwheat groats in a blender to make a coarser flour and voilà, a wonderful new cake was born with the elusive aroma of rose petals and a lovely—not at all gummy—crumb! After that I began to wonder if the difference between our results was also related to mixing. As it turns out, buckwheat flour can turn to mush with excessive mixing—and the finer the flour, the quicker it turns to mush. The butter cake batter is mixed for at least two minutes, while the sponge batter is just gently folded.

Look for a diversity of textures and flavors in this chapter, including gingerbread, spicy pumpkin loaf, date-nut cake, a light sponge cake, fragrant butter cake, linzer cookies, and superb savory crackers. And pay close attention to the mixing instructions; your opinions about buckwheat depend on it.

FLAVOR AFFINITIES FOR BUCKWHEAT FLOUR

Walnuts, toasted hazelnuts, dried fruit, dark spice, coffee, salt, brown sugar, fresh figs, honey

WHERE TO BUY AND HOW TO STORE

Buckwheat flour is available in better supermarkets in the specialty flour section of the baking aisle, or by mail order from Anson Mills or Bob's Red Mill (see Resources, page 351). Buckwheat flour should be stored in an airtight container, away from heat and light, for 2 to 3 months at room temperature, or 6 months in the refrigerator, and up to 12 months in the freezer.

BUCKWHEAT SPONGE CAKE

Buckwheat adds a delicate (and quite surprising) nuance of honey flavor and a slightly floral note, almost like rose petals, to this light and unusual cake. Serve it with Prunes Poached in Coffee and Brandy (page 336) or with strawberries and whipped cream, Rose Whipped Cream (page 345), or Honey Whipped Cream (page 345), garnished with chopped walnuts or Walnut Praline Brittle (page 337). **SERVES 12 TO 14**

½ cup (110 grams) flavorless vegetable oil (such as safflower, corn, or sunflower; see Note)

⅔ cup (85 grams) buckwheat flour

½ cup (80 grams) white rice flour
NOTE: This recipe is not successful with Thai white rice flour.

6 large eggs

1 cup (200 grams) sugar

Rounded ¼ teaspoon salt

EQUIPMENT

Stand mixer with whisk attachment

Sifter or medium-mesh strainer

10-inch tube pan with removable bottom, ungreased (see Note)

NOTES: Oil instead of butter softens the slightly crystalline texture of the flour and (in comparison with butter) lets more of the complex and delicate flavors of the buckwheat shine through.

A tube pan gives the cake a little extra support.

Position a rack in the lower third of the oven and preheat the oven to 325°F.

Pour the vegetable oil into a 6- to 8-cup bowl and set aside.

In a medium bowl, whisk the buckwheat and rice flours together.

Combine the eggs, sugar, and salt in the bowl of the stand mixer and beat with the whisk attachment on high speed for 7 to 8 minutes. The mixture should be light colored and tripled in volume, and you should see well-defined tracks as the whisk spins; when the whisk is lifted, the mixture should fall in a thick, fluffy rope that dissolves slowly on the surface of the batter.

Remove the bowl from the mixer. Sift one-third of the flour mixture over the eggs. Fold with a large rubber spatula until the flour is almost blended into the batter. Repeat with half of the remaining flour. Fold in the rest of the flour. Scrape about a quarter of the batter into the vegetable oil. Fold until the oil is completely blended into the batter. Scrape the oiled batter over the remaining batter and fold just until blended. Scrape the batter into the pan and tilt to level the batter if necessary.

Bake for 40 to 45 minutes, or until the cake shows signs of shrinking from the sides of the pan and a toothpick inserted in the center emerges clean. Set the pan on a rack to cool.

When the cake is cool, run a small spatula around the inside of the pan, pressing against the sides of the pan to avoid tearing the cake, and around the tube. Invert the pan to remove the cake. The cake should be completely cool before filling, frosting, or storing. The cake may be wrapped airtight and stored at room temperature for 2 to 3 days, or frozen for up to 3 months.

BUCKWHEAT CAKE WITH ROSE APPLES

You might imagine that a cake made with coarse buckwheat meal would be quite hearty and rustic. This one has an unexpectedly delicate texture with just a nuance of crunch from the meal and a gentle hint of rose flavor inherent in the buckwheat. Apples poached with rose hips, with floral notes that mirror the flavors in the cake, are a perfect light accompaniment. **SERVES 12 TO 16**

2⅓ cups (360 grams) white rice flour
—OR—
4 cups (360 grams) Thai white rice flour

½ cup plus 2 tablespoons (100 grams) homemade buckwheat meal (see Note)

2 cups minus 3 tablespoons (360 grams) sugar

½ pound (2 sticks/225 grams) unsalted butter, very soft

¾ teaspoon salt

2 teaspoons baking powder

1 teaspoon baking soda

½ teaspoon xanthan gum

1 cup plain yogurt (any percent fat) or slightly watered down Greek yogurt

4 large eggs

2 teaspoons pure vanilla extract

Rose Apples (recipe follows)

1 quart sweet cream or vanilla ice cream, or 2 cups heavy cream, whipped (optional)

EQUIPMENT

Two 9-by-2-inch round cake pans or two 8-inch square pans

Stand mixer with paddle attachment

Position a rack in the lower third of the oven and preheat the oven to 350°F. Grease the sides of the pans with vegetable oil spray or butter and line the bottoms with parchment.

Combine the rice flour, buckwheat meal, sugar, butter, and salt in the bowl of the stand mixer and beat with the paddle attachment on medium speed until the mixture has the texture of brown sugar, about a minute. Add the baking powder, baking soda, xanthan gum, yogurt, eggs, and vanilla and beat on medium-high speed for 2 to 3 minutes; the batter should be very smooth and fluffy.

Scrape the batter into the prepared pans and bake for 25 to 30 minutes for round pans (or about 5 minutes longer for square pans), until a toothpick inserted in the center comes out clean. Cool the cakes in the pans on a rack.

Slide a slim knife or a small metal spatula around the edges of each cake to detach it from the pan. Invert the cakes onto a rack and peel off the paper liner. Turn the cakes right side up.

Serve slices in shallow bowls with several tablespoons of the apples and their syrup. Add a scoop of ice cream or a dollop of whipped cream beside the cake, if desired. This cake keeps, covered airtight, for up to 3 days at room temperature.

NOTE: Pulverize a slightly rounded cup (100 grams) raw or roasted buckwheat groats in a blender to a consistency like sand, not as fine as flour. The coarsely ground buckwheat meal actually makes the texture of this cake light and prevents it from becoming gummy from mixing. See page 168 for more details.

(recipe continues)

Rose Apples

Dried rose hips are available in the bulk bins at health food stores, spice stores, and tea shops. Avoid rose hips from the craft store intended for potpourri! **MAKES 4 CUPS**

¼ cup (20 grams) dried rose hips

1¼ cups water

1¼ cups sugar

2 tablespoons lemon juice

2 pounds (910 grams) firm apples

EQUIPMENT

Strainer

Basket-type coffee filter or paper towel

Place the rose hips in a medium heatproof bowl. Bring the water to a boil in a medium nonreactive saucepan and pour the water over the rose hips. Let steep for 3 minutes. Line a strainer with a coffee filter or paper towel, set over the saucepan, and pour the rose hips and liquid through the strainer; discard the rose hips. Add the sugar and lemon juice and bring to a simmer over medium heat.

Quarter the apples, core, and cut each quarter into 4 cubes. You should have about 4 cups of fruit. Add the apples to the simmering syrup. Simmer for 8 to 10 minutes, or until just tender, turning carefully once or twice. Cover the pan and let cool. Gently transfer the apples and syrup to a bowl or other container, cover, and refrigerate for at least 2 hours or until well chilled before serving. The apples will keep in the refrigerator for at least a week.

BUCKWHEAT GINGERBREAD

Buckwheat adds an earthy flavor and a pleasing edge of bitterness to this moist and medium-spicy gingerbread. I have to keep this cake out of sight to keep myself from "trimming" it, slice by slice, every time I see it. **SERVES 8 TO 10**

¾ cup plus 2 tablespoons (110 grams) brown rice flour

1 teaspoon baking soda

Scant ¼ teaspoon xanthan gum

½ teaspoon ground cinnamon

½ teaspoon ground ginger

¼ teaspoon ground allspice

¼ teaspoon salt

Piece of fresh ginger 1½ by 1½ inches (about 40 grams)

⅔ cup (140 grams) packed light brown sugar

⅓ cup (120 grams) light unsulfured molasses

8 tablespoons (1 stick/115 grams) unsalted butter, melted and still warm

1 large egg

¾ cup plus 2 tablespoons (110 grams) buckwheat flour

½ cup hot water (hot from the tap is fine)

Powdered sugar for dusting and/or lightly sweetened whipped cream (see page 343), Whipped Crème Fraîche (page 345), sour cream, or mascarpone

EQUIPMENT

8-inch round cake pan

Food processor fitted with the steel blade

Position a rack in the lower third of the oven and preheat the oven to 350°F. Grease the sides of the pan with vegetable oil spray or butter and line the bottom with parchment paper.

In a medium bowl, whisk the brown rice flour, baking soda, xanthan gum, cinnamon, ground ginger, allspice, and salt until blended. Set aside.

Peel the ginger with a vegetable peeler. Cut it across the grain into enough thin slices to measure ¼ cup (30 grams). Put the slices in the food processor with the brown sugar and pulse until the ginger is finely pureed with the sugar. Add the molasses, butter, egg, and the reserved flour mixture and process for 15 seconds. Scrape the bowl, add the buckwheat flour and water, and process for just 5 more seconds (overprocessing the buckwheat flour can give the cake a pasty texture).

Scrape the batter into the prepared pan. Bake for 40 to 45 minutes, or until a toothpick inserted in the center comes out clean. Cool the cake in the pan on a rack for 10 minutes.

Run a thin knife or spatula around the sides of the pan to detach the cake. Invert the pan on a rack and peel off the parchment. Turn the cake right side up on the rack to cool completely. Once cool, the cake keeps, wrapped airtight, at room temperature for 3 or 4 days, or in the freezer for up to 3 months; bring to room temperature before serving.

Serve with a dusting of powdered sugar and/or a dollop of whipped cream, whipped crème fraîche, sour cream, or mascarpone.

DATE-NUT CAKE WITH CHERRIES AND BUCKWHEAT

My favorite version of fruitcake is made with nuts and dried (rather than candied) fruit, and it normally has just enough batter to keep it from falling apart. It's rich and inherently sweet from the fruit rather than lots of sugar, which can also be adjusted to your taste. When I tried using buckwheat flour instead of wheat flour in this treasured recipe, the outcome was so flavorful and so delicious with the cherries and walnuts that I ultimately increased the amount of batter just to get even more buckwheat flavor! The result makes a decadent and rather sophisticated snack, a good midafternoon bite with a cup of tea, or a chic addition to a cheese course. See page 244 for a variation with teff flour (photo opposite). **MAKES 1 LOAF, ABOUT 30 VERY THIN SLICES**

½ cup plus 2 tablespoons (70 grams) buckwheat flour

⅛ teaspoon baking soda

⅛ teaspoon baking powder

¼ teaspoon salt

3 to 4 tablespoons (35 to 50 grams) packed light or dark brown sugar

1 cup (175 grams) dates, pitted and cut into quarters

¼ cup (40 grams) lightly packed dried apricot halves or dried pluots, cut in half

¼ cup (40 grams) dried sour or Bing cherries

1¾ cups (175 grams) walnut pieces

2 large eggs

1¼ teaspoons pure vanilla extract

EQUIPMENT

8-by-4-inch (4-cup) loaf pan, bottom and all four sides lined with parchment paper

Position a rack in the lower third of the oven and preheat the oven to 300°F.

In a large bowl, whisk the flour with the baking soda, baking powder, and salt. Add the brown sugar, dates, apricots, cherries, and walnuts and mix thoroughly with your fingers, separating any sticky fruit pieces from one another. Set aside.

Whisk the eggs and vanilla in a medium bowl until lightened in color. Scrape the eggs into the large bowl and mix with a spatula until all of the fruit and nut pieces are coated with batter. Scrape into the prepared pan.

Bake for 60 to 80 minutes, until the top feels firm and crusted. Cool in the pan on a rack. Lift the ends of the paper liner and transfer to a cutting board. Use a sharp knife to cut very thin slices. The cake may be stored, airtight, for at least 1 week at room temperature, or longer in the refrigerator.

DARK AND SPICY PUMPKIN LOAF

Buckwheat flour lends an almost woodsy note to the flavor of this not-too-sweet tea cake. Serve it with coffee, plain or with a smear of cream cheese or soft goat cheese. The batter may also be baked in muffin cups or it may be doubled and baked in a Bundt pan. **SERVES 6 TO 8**

---

8 tablespoons (1 stick/115 grams) unsalted butter, melted

1 cup (200 grams) sugar

2 large eggs

¾ cup (120 grams) white rice flour
—OR—
1¼ cups (120 grams) Thai white rice flour

⅓ cup (40 grams) buckwheat flour

½ teaspoon baking soda

1 teaspoon baking powder

1 teaspoon ground cinnamon

½ teaspoon ground nutmeg

¼ teaspoon salt

¾ cup (170 grams) pumpkin puree

½ cup (70 grams) raisins or currants

EQUIPMENT

8½-by-4½-inch (6-cup) loaf pan, bottom and all four sides lined with parchment paper

Stand mixer with paddle attachment or handheld mixer

Position a rack in the lower third of the oven and preheat the oven to 350°F. Line the bottom and sides of the loaf pan with parchment paper.

Combine the butter, sugar, and eggs in the bowl of the stand mixer and beat on medium speed with the paddle attachment until lighter in color, about 2 minutes. Or beat with the handheld mixer in a large bowl on medium-high speed for 3 to 4 minutes.

Add the rice and buckwheat flours, baking soda, baking powder, cinnamon, nutmeg, salt, pumpkin puree, and raisins and beat on low speed until smooth. Scrape the mixture into the prepared pan.

Bake the loaf for 45 to 50 minutes, until a toothpick or bamboo skewer inserted in the center comes out clean. Cool the loaf in the pan on a rack for at least 2 hours before unmolding and slicing.

The cake keeps, wrapped airtight, in the refrigerator for up to 5 days; let come to room temperature to serve.

(recipe continues)

VARIATIONS

Dark and Spicy Pumpkin Muffins

Line 12 regular muffin cups plus 2 custard cups or ramekins with paper liners (there is too much batter for a 12-muffin tin). Fill each cup about two-thirds full. Bake at 375°F for about 20 minutes, until a toothpick inserted into the center of a muffin comes out clean. Frost with Cream Cheese Frosting (page 102), if desired.

Dark and Spicy Pumpkin Bundt Cake

Spray a 10- to 12-cup Bundt pan with vegetable oil spray. Double the recipe for Dark and Spicy Pumpkin Loaf and bake it in the prepared pan for 50 to 55 minutes, until a toothpick or bamboo skewer inserted in the center comes out clean. Cool the cake in the pan on a rack for 10 to 15 minutes before unmolding it on the rack to cool completely.

BUCKWHEAT COFFEE BABY CAKES WITH TOFFEE SAUCE

I love the dark and intriguing flavors of this surprising combination of buckwheat flour, coffee, and dates. Toffee sauce—not to mention whipped cream—makes almost anything delicious, but it is especially good here. **SERVES 13**

---

1 cup plus 2 tablespoons (150 grams) brown rice flour

⅓ cup plus 1 tablespoon (50 grams) buckwheat flour

8 tablespoons (1 stick/115 grams) unsalted butter, very soft

1 cup (200 grams) sugar

Scant ½ teaspoon salt

1 teaspoon baking powder

½ teaspoon baking soda

¼ teaspoon xanthan gum

½ cup strong regular coffee (or decaf), at room temperature

2 large eggs

2 teaspoons pure vanilla extract

1 cup (170 grams) chopped pitted dates

Toffee Sauce (recipe follows)

1 cup heavy cream, whipped (not sweetened)

EQUIPMENT

12-cup muffin pan

One 6-ounce ramekin or custard cup

Stand mixer with paddle attachment

Position a rack in the lower third of the oven and preheat the oven to 350°F. Line the muffin cups and ramekin (for the extra batter) with paper liners.

Combine the brown rice and buckwheat flours, butter, sugar, and salt in the bowl of the stand mixer and beat on medium speed with the paddle attachment until the mixture is the texture of brown sugar, about a minute. Add the baking powder, baking soda, xanthan gum, coffee, eggs, and vanilla and beat on medium-high speed for 2 to 3 minutes; the batter should be smooth and fluffy and lightened in color. Beat in the dates.

Fill all 13 lined cups about two-thirds full (don't try to fit all of the batter into 12 cups—they will overflow!) and bake for 25 to 30 minutes, until a toothpick inserted in the center of a cake comes out clean. Set the pan and ramekin on a rack to cool.

To serve, pour 2 to 3 tablespoons of Toffee Sauce into shallow bowls. Peel the liners off the baby cakes and place one upside down in each bowl. Top the cakes with a dollop of whipped cream. The cakes keep at room temperature, in an airtight container, for up to 3 days.

(recipe continues)

Toffee Sauce

Toffee sauce is a luscious sauce very simply made with common pantry ingredients. Unrefined sugars such as Demerara, palm sugar, or light muscavado sugar are also wonderful here. Toffee sauce differs from caramel sauce in that the flavor comes from the molasses in the sugar; caramel sauce gets its flavor and color from slightly burnt (caramelized) white sugar. **MAKES 2 GENEROUS CUPS**

1½ cups (300 grams) packed brown sugar

2 cups heavy cream

Pinch of salt

1 tablespoon plus 1 teaspoon pure vanilla extract

Mix the brown sugar and cream in a small saucepan over medium heat. Turn the heat to high and bring the mixture to a boil, stirring constantly. Stir in the salt. Reduce the heat to low and cook for at least 15 more minutes, stirring occasionally, until the sauce is slightly thickened. Remove from the heat and stir in the vanilla.

The sauce may be served immediately, kept warm over low heat to serve later, or cooled and reheated. It keeps in the refrigerator in a covered container for up to a week. Reheat it gently on the stove or (carefully) in a microwave.

PANFORTE NERO

Panforte Nero is the darker and spicier cousin of Siena's renowned fruit and nut confection. The toasty, earthy, almost cocoa-y flavors of buckwheat make it a natural addition. I almost wonder why the bakers of Siena never added it themselves! Long-keeping panforte is the perfect thing to have on hand for afternoon coffee (or espresso) and a luxurious complement to fresh or ripe cheeses. Make two and give one as a gift. **SERVES 12 TO 16**

---

1 cup (135 grams) toasted skinned hazelnuts

¾ cup (105 grams) whole natural (unskinned) almonds, toasted

1¼ cups (150 grams) buckwheat flour

2 tablespoons (12 grams) natural unsweetened cocoa powder

2¼ teaspoons fennel seeds

½ teaspoon ground cinnamon

⅛ teaspoon ground cloves

¼ teaspoon ground black pepper

¼ teaspoon ground ginger

¼ teaspoon ground coriander seeds

¼ teaspoon freshly grated nutmeg

1 tablespoon finely grated orange zest, from an unsprayed orange

1⅓ cups (225 grams/8 ounces) stemmed and halved dried figs

⅔ cup (225 grams) honey

⅔ cup (130 grams) granulated sugar

Powdered sugar for dusting

EQUIPMENT

8-inch round cake pan

Position a rack in the lower third of the oven and preheat the oven to 300°F. Cut a 12-inch circle of parchment paper and place it over an inverted 8-inch cake pan. Pleat the edges of the paper to fit over the sides of the pan. Remove the paper, turn the pan right side up, and fit the paper into the pan; set aside.

In a large bowl, mix the hazelnuts, almonds, buckwheat flour, cocoa, fennel seeds, cinnamon, cloves, pepper, ginger, coriander, nutmeg, orange zest, and figs.

In a large saucepan, mix the honey and granulated sugar and bring to a full boil; boil for 15 seconds. Remove from the heat, add the other ingredients, and mix well.

Scrape the mixture into the center of the prepared pan and spread evenly to the edges. Bake for 50 to 55 minutes, until no longer gooey in the middle. Set the pan on a rack to cool. Lift the edges of the paper liner to remove the panforte. Peel off the liner. Dust the top, bottom, and sides with powdered sugar and wrap airtight in plastic wrap.

To serve, cut thin slices with a thin serrated knife. The panforte keeps, wrapped airtight, at cool room temperature for several months.

BETTER THAN BUCKWHEAT BLINI

These are not American-style pancakes, nor are they the type of buckwheat blini that you may have been served with caviar. Many versions of these crisp-at-the-edges, spongy-and-chewy-in-the-middle pancakes exist around the world. The yeast and yogurt give them a satisfying tanginess. Sweet rice—aka glutinous rice—flour produces a pleasing chewiness in yeast-raised recipes such as this one. No caviar in your fridge right now? Serve them topped with butter and honey for breakfast or tea. **SERVES 8 AS AN APPETIZER**

½ cup plus 2 tablespoons (80 grams) buckwheat flour, or scant ½ cup (80 grams) toasted or raw buckwheat groats

3 tablespoons warm water (105°F to 115°F)

1 teaspoon active dry yeast

½ teaspoon sugar

1 cup plus 1 tablespoon (160 grams) white rice flour
—OR—
1½ cups (160 grams) Thai white rice flour

½ cup (80 grams) glutinous rice flour
—OR—
⅔ cup (80 grams) Thai glutinous white rice flour

2 tablespoons plain yogurt (any percent fat) or slightly watered down Greek yogurt

1 teaspoon salt

1½ cups water

Soybean or corn oil, for frying

½ cup (170 grams) honey

4 tablespoons (½ stick/55 grams) unsalted butter, melted

EQUIPMENT

12-inch frying pan

If using the groats, pulverize them in a blender just to the consistency of sand, not fine flour.

Combine the warm water, yeast, and sugar in a large bowl and let rest for 5 minutes. Add the buckwheat and rice flours, yogurt, salt, and 1½ cups water and whisk until smooth. Cover the bowl with plastic wrap and set over a slightly smaller bowl filled with warm water—the water can touch or not. Let rest for about half an hour. The batter should be fairly thick; if it is not, whisk in a little extra rice flour.

Heat a large frying pan over medium heat, pour in 1 tablespoon oil, and lift and tilt the pan to spread the oil. Whisk or stir the batter to mix thoroughly, and spoon tablespoonfuls of batter into the pan, leaving about an inch between pancakes.

Cook until lightly browned on the first side and bubbles break on the top surface. Use a spatula to turn the pancakes and cook the second side until lightly browned. Repeat with the remaining batter, adding oil to the pan before each batch. Serve immediately or keep warm in a covered dish. To serve, drizzle with honey and butter. Leftovers may be refrigerated for up to 2 days and reheated for 5 minutes in a 400°F oven.

(recipe continues)

VARIATION

Blini with Caviar or Smoked Salmon

For brunch or a light supper for 4 to 6, serve blini with sour cream, onion, and salmon roe or smoked salmon: Have ready 1 cup sour cream, ½ cup (70 grams) finely chopped red onion or several tablespoons of chopped chives, and at least 3½ ounces (100 grams) salmon roe or about 8 ounces (225 grams) smoked salmon. Top each pancake with a small dollop of sour cream, onions, and salmon roe, or let guests serve themselves.

BUCKWHEAT WALNUT OR HAZELNUT TUILES

These crispy cookies make a sophisticated addition to a cookie assortment or an elegant counterpoint to a dish of creamy pudding or vanilla ice cream. Rice appears to be the predominant flour here, but the earthy, nutty flavor of buckwheat is the real star, perfectly partnered with walnuts or hazelnuts—and butter, of course. If you want curved or shaped tuiles, see page 69. **MAKES ABOUT FORTY 3-INCH COOKIES**

---

4 tablespoons (½ stick/55 grams) unsalted butter, melted, plus more unsalted butter for greasing the foil

2 large egg whites

2 teaspoons water

½ cup (100 grams) sugar

2 tablespoons (20 grams) white rice flour
—OR—
3 tablespoons plus 2 teaspoons (20 grams) Thai white rice flour

1 tablespoon plus 1 teaspoon (10 grams) buckwheat flour

Scant ½ teaspoon salt

⅓ cup (35 grams) finely chopped walnuts or toasted skinned hazelnuts

EQUIPMENT

Baking sheets

Silicone baking mats (optional)

Rolling pin or small cups for shaping (optional); see page 69

NOTE: See page 69 for more about pan liners for tuiles.

Line the baking sheets with regular foil (dull side facing up) or silicone baking mats, and grease the foil or silicone lightly but thoroughly. You can also line the pans with nonstick foil (nonstick side up).

In a medium bowl, mix the egg whites with the water, sugar, rice and buckwheat flours, and salt until well blended. Stir in the butter and walnuts. Cover the bowl and let the batter rest for several hours or overnight in the refrigerator to let the flour absorb moisture.

Position racks in the upper and lower thirds of the oven (or one rack in the center if you are baking only one sheet at a time). Preheat the oven to 325°F.

Stir the batter well. Drop level teaspoons 2 inches apart on two lined sheets. Use the back of the spoon to smear the batter into 2½-inch rounds—tuiles baked on greased foil will spread a little; those baked on silicone mats or nonstick foil will not.

Bake, watching carefully, for 12 to 15 minutes, rotating the sheets from front to back and top to bottom about halfway through the baking time, until the tuiles are mostly deep golden brown. If the cookies are not baked enough, they will not be completely crisp when cool, nor will they come off the foil easily.

Slide the foil sheets onto racks and let the cookies cool completely before removing them. Or, if using silicone mats, transfer the hot cookies to a rack to cool.

To retain crispness, put the cookies in an airtight container as soon as they are cool. They will keep in an airtight container for at least 1 month.

BUCKWHEAT SABLÉS

These melt-in-your-mouth cookies have the perfect tender sandy texture for which sablés were named. Serve them with a bowl of fresh blackberries (or other cane berries) and cream. Or turn them into sandwich or linzer cookies (see page 190) filled with a little blackberry or plum preserves or prune butter. These cookies are even better if you can chill the dough overnight before baking. **MAKES ABOUT 3 DOZEN 2-INCH COOKIES**

¼ cup plus 2 tablespoons (55 grams) white rice flour

—OR—

½ cup plus 1 tablespoon (55 grams) Thai white rice flour

½ cup plus 2 tablespoons (70 grams) buckwheat flour

⅔ cup (65 grams) gluten-free oat flour

¼ teaspoon salt

⅛ teaspoon baking soda

½ cup (100 grams) sugar

¼ cup (60 grams) cream cheese, cut into chunks

12 tablespoons (1½ sticks/170 grams) unsalted butter, cut into chunks and softened

1 tablespoon water

EQUIPMENT

Food processor fitted with the steel blade (optional)

Baking sheets, lined with parchment paper

To make the dough by hand, put the rice, buckwheat, and oat flours, salt, baking soda, and sugar in a large bowl and whisk until thoroughly blended. Add the cream cheese, butter, and water. Use a fork or the back of a large spoon to mash and mix the ingredients together until all are blended into a smooth, soft dough.

To make the dough in a food processor, combine the rice, buckwheat, and oat flours, salt, baking soda, and sugar. Pulse to mix thoroughly. Add the cream cheese, butter, and water. Process just until the mixture forms a ball of smooth, soft dough. Scrape the bowl and blend in any stray flour at the bottom with your fingers.

Scrape the dough onto a sheet of wax paper and form it into a 10-inch log about 1¾ inches in diameter. Wrap tightly in the wax paper and refrigerate for at least 2 hours, but preferably longer or overnight.

Position racks in the upper and lower thirds of the oven and preheat the oven to 325°F.

Use a sharp knife to cut the cold logs of dough into ¼-inch slices. Place the cookies at least 1½ inches apart on the prepared baking sheets. Bake for 20 to 25 minutes, rotating the sheets from front to back and top to bottom about halfway through the baking time, until the cookies are slightly darker brown at the edges and well browned on the bottom.

Set the pans or just the liners on racks to cool. Cool completely before stacking or storing. The cookies may be stored in an airtight container for at least 2 weeks.

BUCKWHEAT LINZER COOKIES

These pretty cookies look as though they are fussy to make, but they are actually slice-and-bake cookies, with holes cut from half of them about halfway through the baking. Buckwheat pairs well with any dark berry or cherry flavor, so feel free to try different preserves. The cookies keep well, but they should be assembled only shortly before serving. Leftover filled cookies will soften a bit, but they will still taste great. **MAKES ABOUT 1½ DOZEN 2-INCH SANDWICH COOKIES**

Buckwheat Sablés dough (page 189), shaped into a log and chilled as directed

½ cup blackberry (or other) preserves

Powdered sugar for dusting

EQUIPMENT

Baking sheets, lined with parchment paper

⅞-inch round cookie cutter (or bottle cap to improvise)

Fine-mesh strainer

Position racks in the upper and lower thirds of the oven and preheat the oven to 325°F.

Slice the chilled log less than ¼ inch thick and place the slices 1½ inches apart on the prepared baking sheets, dividing the total number equally between them. Bake for about 12 minutes. Remove the upper sheet of cookies and place it on the counter or stovetop. Press the cookie cutter gently into each cookie. If the centers lift out, fine; otherwise you can remove them later. Switch and rotate sheets, placing the first on the lower rack in place of the second. Bake for 10 to 15 minutes, or until the cookies are slightly darker at the edges and well browned on the bottom.

Set the pans or just the liners on racks to cool. Cool completely. Remove the cutouts. Unfilled cookies may be stored in an airtight container for at least 2 weeks.

Shortly before serving, spread ½ teaspoon of preserves on the cookies without holes. Sieve a little powdered sugar over the cookies with holes and place one on top of each jam-topped cookie.

BUCKWHEAT SOUR CREAM SOUFFLÉS WITH HONEY

Whether or not you love traditional robust buckwheat dishes, you will be delighted by the delicacy of this dessert with its gentle hints of rose and green tea. The starch in the buckwheat flour keeps the eggs from overcooking and provides a slightly creamy texture to the soufflé, as would any other flour or starch in any soufflé, but here, the buckwheat flour is also the main source of flavor, which is what makes these soufflés so exciting and interesting. The best accompaniment is a drizzle of excellent raw honey—buckwheat, orange blossom, or a mixed-flower honey. **SERVES 6**

Softened or melted unsalted butter and sugar for coating the ramekins

1 cup sour cream

3 large eggs, separated

3 tablespoons (25 grams) buckwheat flour

⅛ teaspoon salt

1 teaspoon pure vanilla extract

⅛ teaspoon cream of tartar

5 tablespoons (60 grams) granulated sugar

Powdered sugar for dusting

Buckwheat honey or a floral honey such as orange blossom

EQUIPMENT

Six 6-ounce ramekins

Rimmed baking sheet

Stand mixer fitted with the whisk attachment or handheld mixer

Position a rack in the center of the oven and preheat the oven to 375°F. Butter and sugar the ramekins and place them on the baking sheet.

Combine the sour cream, egg yolks, flour, salt, and vanilla in a large bowl and stir until blended.

Combine the egg whites with the cream of tartar in the bowl of the stand mixer fitted with the whisk attachment (or in another bowl if using a handheld mixer). Beat on medium speed (or on high speed with the handheld mixer) until the egg whites are creamy white and hold a soft shape when the beaters are lifted. Gradually add the granulated sugar, continuing to beat on medium speed until the egg whites are stiff but not dry. Fold about one-quarter of the egg whites into the sour cream mixture, then fold in the remaining egg whites.

Divide the batter among the ramekins, filling them nearly to the top. (The unbaked soufflés can be covered and refrigerated for up to 24 hours.)

Bake the soufflés for 15 to 18 minutes (or a couple of minutes longer if the soufflés have been chilled), until puffed and slightly golden brown. Dust lightly with powdered sugar and serve immediately, passing a dish of honey for drizzling.

WALNUT AND BUCKWHEAT CRACKERS

Dark and crunchy and pleasingly bitter, these crackers are fabulous with a smear of whole-milk Greek yogurt or sour cream topped with lox or salmon roe (or real caviar) and very thinly sliced red onions or shallots. Like the Seed Crackers (page 163) and Tangy Aromatic Crackers (page 259; photo opposite), they keep well and are fantastic to package in cellophane bags and give as a hostess gift. All of these recipes can be doubled or tripled with ease. **MAKES 1½ DOZEN LARGE CRACKERS**

---

1 cup plus 2 tablespoons (100 grams) brown rice flour

¾ cup (120 grams) white rice flour
—OR—
1¼ cups (120 grams) Thai white rice flour

½ cup (60 grams) buckwheat flour

1 cup (100 grams) walnut pieces

¼ cup plus 2 tablespoons (40 grams) flaxseed meal, or ¼ cup (40 grams) whole flaxseeds, finely ground

1 tablespoon packed brown sugar

1½ teaspoons salt

¾ cup plus 1 tablespoon water

1 tablespoon rice vinegar

2 teaspoons baking powder

¼ cup flavorless vegetable oil (such as soybean, corn, or safflower)

EQUIPMENT

Stand mixer with paddle attachment

Rolling pin

Baking sheets

Position racks in the upper and lower thirds of the oven and preheat the oven to 450°F.

Mix the rice and buckwheat flours, walnuts, flaxseed meal, brown sugar, and salt in the bowl of the stand mixer fitted with the paddle attachment. Add the water and vinegar and beat for 2 minutes on medium speed to form a very thick, sticky dough that might wrap around the paddle at first. Sprinkle in the baking powder, add the oil, and beat for 1 minute on medium speed to thoroughly incorporate the oil.

Cut four pieces of parchment the size of a baking sheet. Drop three 2-tablespoon lumps of dough evenly spaced down the length of a parchment sheet. Cover with another piece of parchment and flatten each lump with the heel of your hand. Use a rolling pin to roll the dough into oblongs about 3 by 5 inches to about 4 by 8 inches (for even thinner crackers) and a scant ⅛ inch thick. Peel off the top parchment (save for reuse) and place the parchment with the crackers *dough side down* on a baking sheet.

Bake for 5 to 6 minutes, two pans at a time, rotating them from upper to lower and front to back for even baking, until the crackers are browned at the edges. Remove the pan from the oven and carefully peel off the parchment (save for reuse). Flip the crackers over with a spatula and return them to the oven for 2 to 3 minutes, or until well browned at the edges. Repeat with the remaining dough: while the crackers are baking, continue to roll out more dough; as soon as the pans of crackers are done, flip the next batch onto the baking sheets (it's okay if the sheets are still hot as long as you put them in the oven immediately).

Cool the crackers thoroughly on parchment or a rack before storing in an airtight container for up to 10 days. The crackers may be refreshed before serving by baking for 5 minutes at 400°F.

CHAPTER FIVE

CHESTNUT FLOUR

CHESTNUTS were a critically important food in Europe and Asia long before the cultivation of wheat or potatoes. European settlers found an abundance of the trees on American shores, too, dispersed over 200 million acres of eastern woodland, the length of the eastern seaboard, and as far west as the Mississippi. Like their Asian and European counterparts, Native Americans subsisted on chestnuts, raw or roasted, in gruels and stews, and pounded into meal to make unleavened breads, and the new settlers followed suit. Chestnuts might be a staple of our diet today had not a lethal blight wiped out an estimated 4 billion trees—one-quarter of the hardwood forest—in the first half of the twentieth century.

Historical reliance on chestnuts as a subsistence food—and even their use in modern wartime Europe when grain crops and milled flour were scarce—branded chestnuts as the food of the poor. Even today, chestnut trees are called "bread trees" in parts of southern Europe where chestnut flour is still widely used to make bread.

But in affluent countries, especially in urban centers, the food of the poor becomes chic. Chestnuts have become a celebration food. We savor them roasted from carts on snowy street corners and stuff them into holiday birds. Candied chestnuts, chestnut cream, chestnuts in syrup, chestnut honey, and vacuum-packed steamed chestnuts are luxe items sold in gourmet shops, where, with luck, you might also find chestnut flour.

Chestnut flour is milled from pulverized dried chestnuts and has a soft, starchy texture and a tendency to clump, similar to oat flour. 78 percent carbohydrates and only 1 percent fat. It varies in color from pale to warm tan—sometimes related to whether the flour is raw or roasted—and has a sweet aroma and a sweet, slightly nutty flavor. Raw and roasted flours are interchangeable in recipes, but I prefer the flavor of raw flour. Either way, beware of flour that smells and tastes excessively smoky. It is traditional in Italy to dry the nuts over a wood fire, but the flavor of some flours produced this way is decidedly tainted by too much smoke. For desserts, especially, you'll want to use flour with little if any smoky flavor.

Chestnut flour gives cakes a very soft crumb that is never gritty and has plenty of flavor. It can be used alone, without any other flour, and it works wonderfully in all kinds of sponge and egg-based cakes, meringues, and even a simple homey pudding. Paired with rice flour, it makes a flavorful shortbread crust for tarts (see page 209)

FLAVOR AFFINITIES FOR CHESTNUT FLOUR

Honey (especially chestnut honey), hazelnuts, walnuts, pine nuts dark or white chocolate, brandy or cognac, Grand Marnier, sweet fortified wine such as sherry or marsala, caramel, cardamom, fresh cheese such as ricotta, mascarpone cheese, crème fraîche, cinnamon, coffee, figs, ginger, maple, prunes and plums, pears, apples, orange zest, brown sugar

WHERE TO BUY AND HOW TO STORE

Chestnut flour is available in some better supermarkets and specialty stores, sometimes only during the fall holidays. It is otherwise available by mail order, year-round. If you are not using it up within 2 to 3 months, keep chestnut flour in an airtight container in the refrigerator or freezer. See Resources (page 351) for my preferred source.

CHESTNUT SPONGE CAKE WITH PEAR BUTTER AND CRÈME FRAÎCHE

Layers of chestnut cake spread with a little spiced pear butter (you could also use apple butter) and filled with a thick layer of whipped crème fraîche make for a very simple yet sophisticated dessert. **SERVES 10 TO 12**

FOR THE CAKE

3 tablespoons (45 grams) Clarified Butter (page 335) or ghee (see page 38)

1 cup (100 grams) chestnut flour

⅔ cup (130 grams) sugar

4 large eggs

⅛ teaspoon salt

FOR THE FILLING

1 cup crème fraîche, cold

1 teaspoon pure vanilla extract

2 to 3 teaspoons granulated sugar

⅓ to ½ cup purchased pear or apple butter

Powdered sugar for dusting

EQUIPMENT

8-by-2-inch round cake pan

Stand mixer with whisk attachment

Sifter or medium-mesh strainer

Position a rack in the lower third of the oven and preheat the oven to 350°F. Line the bottom of the pan with parchment paper, leaving the sides ungreased.

Put the clarified butter in a small pot or microwavable container, ready to reheat when needed, and have a 4- to 5-cup bowl ready to pour it into as well—the bowl must be big enough to allow you to fold some batter into the butter later.

Whisk the flour and 2 tablespoons of the sugar together thoroughly in a medium bowl.

Combine the remaining sugar, eggs, and salt in the bowl of the stand mixer and beat with the whisk attachment on high speed for 4 to 5 minutes. The mixture should be light colored and tripled in volume, and you should see well-defined tracks as the whisk spins; when the whisk is lifted, the mixture should fall in a thick, fluffy rope that dissolves slowly on the surface of the batter.

Just before the eggs are ready, heat the clarified butter until very hot and pour it into the reserved bowl.

Remove the bowl from the mixer. Sift one-third of the flour over the eggs. Fold with a large rubber spatula until the flour is almost blended into the batter. Repeat with half of the remaining flour. Repeat with the rest of the flour. Scrape about one-quarter of the batter into the bowl of hot butter. Fold until blended. Scrape the buttery batter over the remaining batter and fold just until blended. Scrape the batter into the pan.

Bake for 25 to 30 minutes, until the cake is golden brown on top, just barely shows signs of shrinking from the sides of the pan, and a toothpick inserted in the center comes out clean. Set the pan on a rack. While the cake is still hot, run a small spatula around the inside of the pan, pressing against the sides of the pan to avoid tearing the cake.

At your convenience (the cake can be warm or completely cool), invert the pan onto a rack to remove the cake and peel off the parchment liner. Turn the cake right side up to finish cooling. The cake should be completely cool before filling, frosting, or storing. The cake may be wrapped airtight and stored at room temperature for 2 or 3 days, or frozen for up to 3 months.

To finish the cake, whip the crème fraîche and vanilla in a chilled bowl with chilled beaters until it begins to thicken. Add granulated sugar to taste, beating until the cream holds a good shape without being too stiff—it will continue to stiffen as you spread it on the cake. Refrigerate the cream while you prepare the next steps.

Turn the cake best-looking side up on a platter. Cut the cake into 2 layers with a serrated bread knife. If the top layer is too delicate to pick up without breaking, slide a rimless baking sheet or a flexible plastic cutting mat under it and set it aside. Spread the bottom layer evenly with the pear butter, then all of the whipped crème fraîche. Set the top cake layer on top of the cream and press gently with the bottom of a pan to level the cake. Cover the cake and refrigerate it for at least 2 hours, and up to 1 day.

Sieve a little powdered sugar over the top of the cake before serving. Leftovers keep in an airtight container in the refrigerator for another day or so.

VARIATIONS

Chestnut Sponge Cake with Figs

Substitute ⅓ to ½ cup fig preserves for the pear butter and 1 cup heavy cream for the crème fraîche. If figs are in season, serve cake slices with quartered fresh ripe figs.

Chestnut Sponge Cake with Praline Whipped Cream

Omit the pear butter and substitute Praline Whipped Cream (made with walnuts or toasted hazelnuts, page 345) for the crème fraîche.

CHESTNUT BÛCHE DE NOËL

A homemade bûche de Noël is the most stunning and festive holiday centerpiece I know. My favorite was always plain génoise filled with chestnut buttercream and finished with chocolate buttercream and meringue mushrooms. Here, a flavorful caramel-hued chestnut sponge cake brushed with a hint of rum or bourbon syrup takes the place of the plain génoise, and the meringue mushrooms are made with chestnut flour. If you prefer not to use alcohol, the cake is moist and flavorful enough to simply omit the syrup. Several tricks—honed over my fourteen years as the owner of a bakery—include freezing the cake before frosting it and cutting the "stumps," using a plastic fork to texture the buttercream, and using two dinner forks to lift the cake onto the platter without breaking the cake or messing up the frosting. **SERVES 12 TO 14**

FOR THE CAKE

4 tablespoons (55 grams) Clarified Butter (page 335) or ghee (see page 38)

1¼ cups (125 grams) chestnut flour

¾ cup (150 grams) sugar

5 large eggs

Generous ⅛ teaspoon salt

FOR THE SYRUP

2 tablespoons sugar

2 tablespoons bourbon, brandy, or rum

1 tablespoon plus 2 teaspoons water

FOR THE FILLING AND FROSTING

7 ounces (200 grams) 55% to 62% dark chocolate, coarsely chopped

¼ cup water

2 cups New Classic Buttercream (page 348)

½ cup chestnut spread (sweetened chestnut puree), or more to taste

Chestnut Meringue Mushrooms (page 219) for decoration (optional)

A few raw cranberries rolled in sugar for decoration (optional)

Position a rack in the center of the oven and preheat the oven to 350°F. Line the bottom of the pan with parchment paper.

Put the clarified butter in a small pot or microwavable container ready to reheat when needed, and have a 4- to 5-cup bowl ready to pour it into as well—the bowl must be big enough to allow you to fold some batter into the butter later.

Whisk the flour and 2 tablespoons of the sugar together thoroughly in a medium bowl.

In the stand mixer fitted with the whisk attachment, beat the eggs, remaining sugar, and salt at high speed for 5 to 6 minutes. The mixture should be light colored and tripled in volume, and you should see well-defined tracks as the whisk spins; when the whisk is lifted, the mixture should fall in a thick, fluffy rope that dissolves slowly on the surface of the batter.

Just before the eggs are ready, heat the clarified butter until it is very hot and pour it into the reserved bowl.

Remove the bowl from the mixer. Sift one-third of the flour over the eggs. Fold with a large rubber spatula until the flour is almost blended into the batter. Repeat with half of the remaining flour, then fold in the rest of the flour. Scrape about a quarter of the batter into the bowl of hot butter. Fold until blended. Scrape the buttery batter over the remaining batter and fold just until blended. Scrape the batter into the pan and spread it evenly with an offset spatula, using as few strokes as possible to avoid deflating the batter.

EQUIPMENT

16-by-12-by-1-inch half sheet pan or 11-by-17-inch jelly roll pan

Stand mixer with whisk attachment

Sifter or medium-mesh strainer

Bake for 10 to 15 minutes, or until the top is golden brown and springs back when pressed gently with your fingers. Set the pan on a rack to cool completely before filling.

Meanwhile, make the syrup by combining the sugar, spirits, and water in a small jar or bowl. Stir, cover, and let stand to dissolve the sugar, at least 15 minutes or until needed.

Run a knife around the edges of the pan to detach the cake. Cover the cake with a sheet of wax paper and set a baking sheet on top, hold the pan and sheet together, and flip them over. Remove the pan and peel the parchment off the sponge. Cover the cake with a sheet of foil, top with the pan and flip the whole business over again, and remove the baking sheet. The cake should be right side up on the foil, in the pan. Use the foil to lift or slide the cake onto the counter.

Brush the cake with the syrup to moisten it slightly.

Place the chocolate and water in a stainless steel bowl and set it directly in a wide skillet of barely simmering water. Stir frequently until the chocolate is melted and the mixture is smooth. Let the chocolate mixture cool to lukewarm and stir it into 1 cup of the buttercream. Cover and refrigerate the chocolate buttercream (or freeze it in an airtight container) until you are ready to frost the cake.

Stir the chestnut spread into the remaining cup of buttercream.

Spread the chestnut buttercream evenly over the cake. Start rolling the cake at one short end by folding the edge about ½ inch over the filling and continue to roll the cake, using the foil beneath it to help. Roll the cake gently but tightly, as though it were a sleeping bag, keeping the roll as cylindrical as possible. When the roll is complete, wrap it tightly in foil and freeze until hard, or at least firm (the firmer the cake, the easier it will be to decorate). The cake may be completed to this point, and kept frozen for up to 2 months.

To finish the cake, soften and stir the reserved chocolate buttercream, as directed on page 349, until it is smooth and creamy. It should be very soft so it doesn't set immediately when you spread it on the frozen cake. If necessary, set the bowl in warm water and stir for just a few extra seconds, until the consistency is right.

(recipe continues)

Unwrap the cake and trim a fraction of an inch from each end to even them. Set the cake on a decorating turntable or a baking sheet. Have a plastic fork (if possible) ready to texture the buttercream and two regular dinner forks ready to transfer the bûche to a serving platter.

Use a spatula to spread the log lavishly with chocolate buttercream, reserving at least a couple of tablespoons. Immediately (before the buttercream starts to stiffen) rake the plastic fork through it so that it resembles the texture of tree bark. Dip a large sharp knife in hot water, wipe it dry, and cut a "stump," at a slant, from each end of the log. Set the stumps aside. Slip the dinner forks under the log—one near each end—and then lift it and transfer it to a serving platter. Position the stumps against the cake on either side, slanted sides facing out. Cover any bare spots and gaps between the stumps and the cake with the reserved buttercream and use the plastic fork to touch up the "bark" effect as necessary.

If you like, decorate the cake with fresh sprigs of greenery, meringue mushrooms, and sugared cranberries. Unless you have a covered container that will fit it, refrigerate the cake uncovered until the buttercream is set, and then wrap the whole platter gently with plastic wrap. For the best flavor, texture, and appearance, remove the cake from the refrigerator (and remove the plastic wrap) at least an hour before serving. The cake keeps, covered and refrigerated, for 3 or 4 days.

CHOCOLATE CHESTNUT SOUFFLÉ CAKE

Chestnut flour adds a delicate flavor and a soft texture to this otherwise flourless chocolate torte. The trick to getting this torte just right is using a dark chocolate with a cacao percentage that is not so high that it overwhelms the chestnut flavor. For extra flavor and a little crunch, you can sprinkle the top of the batter with pine nuts and let them toast as the torte bakes. Serve this cake with plain or coffee whipped cream (see pages 343 and 345). For the best flavor, bake the cake a day before serving. **SERVES 10 TO 12**

---

4 ounces (115 grams) 60% to 64% chocolate, coarsely chopped

8 tablespoons (1 stick/115 grams) unsalted butter, cut into chunks

4 large eggs, separated, at room temperature

½ cup (100 grams) granulated sugar

⅛ teaspoon salt

¼ teaspoon cream of tartar

½ cup (55 grams) chestnut flour

3 to 4 tablespoons raw or lightly toasted pine nuts (optional)

Powdered sugar for dusting (optional)

EQUIPMENT

Stand mixer with whisk attachment, or handheld mixer

8-by-3-inch springform pan or cheesecake pan with removable bottom, ungreased

Position a rack in the lower third of the oven and preheat the oven to 375°F.

Melt the chocolate and butter in a stainless steel bowl set directly in a wide skillet of barely simmering water, stirring occasionally until the chocolate is almost completely melted. Remove from the heat and stir until the chocolate is completely melted and the mixture is smooth.

In a large bowl, whisk the egg yolks with half of the granulated sugar and the salt until pale and thick. Stir in the warm chocolate. Set aside.

Combine the egg whites and cream of tartar in the bowl of the stand mixer fitted with the whisk attachment (or in another large bowl if using a handheld mixer). Beat on medium speed (or on high speed with the handheld mixer) until the egg whites are creamy white and hold a soft shape when the beaters are lifted. Gradually sprinkle in the remaining granulated sugar, beating on high speed until the eggs whites are stiff but not dry.

Immediately whisk the chestnut flour into the chocolate batter, then scrape about one-fourth of the egg whites on top of the chocolate batter and fold just until blended. Fold in the remaining egg whites. Scrape the batter into the pan and spread to level if necessary. Sprinkle the batter with the pine nuts, if using.

Bake for 25 to 30 minutes, until a toothpick inserted in the center of the cake comes out almost clean. Set the pan on a rack to cool completely. The cake will sink and may crack on the surface as it cools.

Slide a slim knife around the sides of the cake to detach it. Remove the pan sides and transfer the cake to a serving platter. The torte may be kept at room temperature, covered or under a cake dome, for at least 3 days. Sprinkle a little powdered sugar over the top before serving if desired.

WALNUT AND HONEY TART WITH CHESTNUT CRUST

The original Engadiner Nusstorte is an old-school Swiss classic: a double-crusted tart filled with sumptuously rich honey caramel loaded with walnuts. Think of it being served for afternoon tea in a decadent old-world café. The new Engadiner, with a chestnut flour crust, is even more flavorful. This confection keeps very well without refrigeration and travels well. Thin slices are perfect finger food for a holiday sweets table. **MAKES 16 THIN SLICES OR ABOUT 24 EXTRA-THIN SLICES**

FOR THE CRUST

1½ cups (150 grams) chestnut flour

⅓ cup plus 1 tablespoon (60 grams) white rice flour
—OR—
½ cup plus 1 tablespoon (60 grams) Thai white rice flour

¼ cup plus 2 tablespoons (75 grams) sugar

Scant ½ teaspoon salt

12 tablespoons (1½ sticks/170 grams) unsalted butter, slightly softened and cut into chunks

¼ cup (60 grams) cream cheese

1 egg mixed with 1 tablespoon milk and a pinch of salt, for the egg wash

FOR THE FILLING

1 cup (200 grams) sugar

⅓ cup (115 grams) honey

2 tablespoons light corn syrup

½ teaspoon lemon juice

⅛ teaspoon salt

1 cup heavy cream

3 tablespoons (45 grams) unsalted butter, cut into chunks

1½ teaspoons pure vanilla extract

2¾ cups (275 grams) broken or very coarsely chopped walnuts

Grease the tart pan with vegetable oil spray or butter.

To make the crust by hand, put the chestnut and rice flours, sugar, and salt in a large bowl and whisk until thoroughly blended. Add the butter chunks and cream cheese. Use a fork or the back of a large spoon to mash and mix the ingredients together until all are blended into a smooth, soft dough.

To make the crust in a food processor, put the chestnut and rice flours, sugar, and salt in the food processor. Pulse to blend. Add the butter chunks and cream cheese. Pulse until the mixture forms a smooth, soft dough. Scrape the bowl and blend in any stray flour at the bottom of the bowl with your fingers.

Divide the dough into two unequal portions, about 60/40, and shape each into a flat disk.

Roll the smaller piece gently between two sheets of wax paper to a circle about 10 inches in diameter and ⅛ inch thick. Without removing the wax paper, slide the circle onto a baking sheet and refrigerate. Line the tart pan with the remaining dough: Use the heel of your hand and then your fingers and/or a small offset spatula to flatten the dough all over the bottom of the pan. Press it squarely into the corners of the pan with the side of your index finger to prevent extra thickness at the bottom edges, and press it as evenly as possible up the sides of the pan, extending about ¼ inch above the rim. Have patience; there is just enough dough (although you may not think so at first). If there is too much dough in one place (or hiding in the corners of the pan), pinch or scrape it off and move it elsewhere. Spread or smear it smooth with the spatula. Here's a final trick for a perfectly even crust: Press a sheet of plastic wrap against the bottom and up the sides of the pan and

EQUIPMENT

Food processor fitted with the steel blade (optional)

Rolling pin

9½-inch tart pan with a removable bottom

Rimmed baking sheet

Candy thermometer

lay a paper towel on top. Set a straight-sided, flat-bottomed cup on the towel; press and slide the cup all over the bottom and into the corners to smooth and even the surface. Leave the plastic wrap in place. Refrigerate the pan for at least 2 hours, but preferably overnight and up to 3 days.

To make the filling, in a 1½- to 2-quart heavy-bottomed saucepan, combine the sugar, honey, corn syrup, lemon juice, and salt. Cook over medium heat, stirring gently with a silicone spatula or wooden spoon, until the mixture is liquefied and begins to simmer around the edges. Wipe the sugar and syrup from the sides of the pot with a wet pastry brush or wad of paper towel. Cover and continue to cook for about 3 minutes. Rinse the sugar crystals from the spatula or spoon before using it again later. Uncover the pot and wipe the sides again. Insert a candy thermometer without letting it touch the bottom of the pot. Cook uncovered, without stirring, until the mixture reaches 305°F. Meanwhile, heat the cream to a simmer in a small pot. Turn off the burner and keep the hot cream handy.

As soon as the sugar mixture reaches 305°F, turn off the heat. Immediately stir in the butter chunks. Gradually stir in the hot cream; it will bubble up and steam dramatically, so be careful. Turn on the burner and adjust the heat so that the mixture boils energetically but not violently. Continue to cook, stirring occasionally, until the thermometer registers 246°F.

Meanwhile, position a rack in the lower third of the oven and preheat the oven to 350°F. Remove the tart pan from the refrigerator and peel off the plastic wrap.

When the syrup is done, remove the pot from the heat. Stir in the vanilla and walnuts. Scrape the filling into the lined pan and spread it evenly. Wait a minute or two for the top edge of the dough to soften so you can press it down, level with the rim of the pan, forming a thicker top edge to attach the top crust. Use your fingers to moisten the edge with a little of the egg wash, without getting it on the rim of the pan itself. Take the pastry circle from the refrigerator and peel off the top sheet of paper. Lift the edges of the bottom sheet of paper and invert the pastry over the tart. Press well to seal the pastry at the edges and trim the excess against the rim of the pan. Remove the paper and trimmings. Brush the top with a thin coat of egg wash.

Set the pan on the rimmed baking sheet and bake for 15 minutes. Slide the oven rack out and use a sharp knife to cut a large X into the top crust: Start about 2 inches from one edge and cut across to the other side, rotate the pan 90 degrees, and repeat. Bake for 20 to 30 more minutes to ensure that the bottom crust is well browned. If the top is getting too brown too fast, lay a sheet of foil loosely on top.

Set the pan on a rack to cool for 20 to 30 minutes, then loosen or remove the rim of the pan while the tart is still warm to avoid sticking. Cool for at least 4 hours or overnight before serving.

Store and serve at room temperature. Wrapped airtight, the tart keeps (some say it improves with age) for at least a week.

When a Tart Sticks to the Edge of the Pan

Any tart with filling that can leak or get sticky around the rim of the pan should be detached from the pan while it is still warm, to prevent it from becoming cemented to the pan when cool. Here, the tart is cooled just long enough for the crust to firm up and then loosened before it is cooled completely. Occasionally, a tart will stick despite your best efforts. Don't panic. Carefully slip the point of a small knife or the curved tip of an old-fashioned vegetable peeler in each flute, between the crust and the pan, wherever there is sticking. It happens to the best of us.

CHESTNUT JAM TART

A jam tart seems like a relaxed, simpler-to-make linzer torte, with an Italian accent instead of a German one. A jam tart is called *fregolata* in Italian, and it's pretty and festive and giftable, too. I thought it fitting (and extra delicious) to swap the usual shortbread crust for a chestnut crust. The dough is quick to make by hand and is then pressed flat into a tart pan with no worries about the sides since the dough forms its own edge as it bakes. Any jam will do for the topping, but the prettiest and most flavorful are red fruits like cherry, plum, raspberry, blackberry, or even strawberry. The jam is topped with crumbled bits of dough and sliced almonds and pushed into the oven to do its own thing. **MAKES 10 SERVINGS**

1½ cups (150 grams) chestnut flour

⅓ cup plus 1 tablespoon (60 grams) white rice flour

—OR—

½ cup plus 1 tablespoon (60 grams) Thai white rice flour

½ cup (100 grams) sugar

Rounded ¼ teaspoon salt

⅛ teaspoon baking soda

12 tablespoons (1½ sticks/170 grams) unsalted butter, slightly softened, cut into chunks

¼ cup (60 grams) cream cheese

⅓ cup (100 grams) jam or preserves

¼ cup (25 grams) sliced almonds

EQUIPMENT

9½-inch fluted tart pan with a removable bottom

Rimmed baking sheet

To make the crust, put the chestnut and rice flours, sugar, salt, and baking soda in a large bowl and whisk until thoroughly blended. Add the butter chunks and cream cheese. Use a fork or the back of a large spoon to mash and mix the ingredients together until all are blended into a smooth, soft dough.

Set aside a scant ¼ cup of the dough. Press and spread the remaining dough evenly across the bottom of the tart pan, but not up the sides. Wrap and refrigerate the pan and the reserved piece of dough for at least 2 hours and up to 3 days.

Position a rack in the lower third of the oven and preheat the oven to 325°F.

Spread the jam evenly over the dough, leaving a scant ½-inch border all around. Without completely covering the jam—because it's pretty—sprinkle it with crumbled bits of the reserved dough and the almonds.

Set the pan on the baking sheet. Bake 35 to 40 minutes, until the edges are rich golden brown and have pulled away from the sides of the pan. If the tart puffs up during baking—take a peek after about 20 minutes, and a couple of times thereafter—settle it down by lifting the edge of the baking sheet and rapping it sharply on the oven rack a couple of times. Cool the tart in the pan on a rack.

When the tart is cool, remove the rim of the pan and set the tart on a serving plate. The tart keeps, covered, at room temperature for at least 3 days.

RICOTTA CHEESECAKE WITH CHESTNUT CRUST

This festive Italian-style cheesecake—now with a chestnut crust—is not as sweet as American cheesecake, and it's all about the ricotta, so put away that silver package of cream cheese. If you have a favorite cheese shop, ask for the best ricotta they have; this cake is worth it. Choose golden raisins rather than candied peel unless you candy your own citrus peels or have an excellent source (the product available in supermarkets at holiday time is awful). Be sure to make the cake at least 2 days ahead. **SERVES 12**

FOR THE CRUST

1 cup plus 2 tablespoons (115 grams) chestnut flour

¼ cup (40 grams) white rice flour
—OR—
⅓ cup plus 1 tablespoon (40 grams) Thai white rice flour

¼ cup plus 2 tablespoons (75 grams) sugar

Scant ½ teaspoon salt

9 tablespoons (130 grams) unsalted butter, slightly softened and cut into chunks

3 tablespoons (45 grams) cream cheese

1 egg yolk mixed with a pinch of salt and ½ teaspoon water, for the egg wash

FOR THE FILLING

3 cups (665 grams) whole-milk ricotta cheese, at room temperature

¾ cup (150 grams) sugar

1 tablespoon white rice flour

1½ teaspoons pure vanilla extract

4 large eggs, at room temperature

2 tablespoons chopped candied orange or lemon peel or golden raisins

2 tablespoons slivered almonds, toasted

¼ cup (30 grams) pine nuts, toasted

To make the crust by hand, put the chestnut and rice flours, sugar, and salt in a large bowl and whisk until thoroughly blended. Add the butter chunks and cream cheese. Use a fork or the back of a large spoon to mash and mix the ingredients together until all are blended into a smooth, soft dough.

To make the crust in a food processor, put the chestnut and rice flours, sugar, and salt in the food processor. Pulse to blend. Add the butter chunks and cream cheese. Pulse until the mixture forms a smooth, soft dough. Scrape the bowl and blend in any stray flour at the bottom of the bowl with your fingers.

The dough may seem much softer than other tart doughs. Use the heel of your hand and then your fingers and/or a small offset spatula to spread the dough all over the bottom of the pan. Press it squarely into the corners of the pan with the side of your index finger to prevent extra thickness at the bottom edges, and press it as evenly as possible about halfway up the sides of the pan. Have patience; there is just enough dough (although you may not think so at first). If there is too much dough in one place (or hiding in the corners of the pan), pinch or scrape it off and move it elsewhere. Spread or smear it smooth with the spatula. Here's a final trick for a perfectly even crust: Press a sheet of plastic wrap against the bottom and up the sides of the pan and lay a paper towel on top. Set a straight-sided, flat-bottomed cup on the towel; press and slide the cup all over the bottom and around the sides to smooth and even the surface. Leave the plastic wrap in place. Refrigerate the pan for at least 2 hours, but preferably overnight and up to 3 days.

Position a rack in the lower third of the oven and preheat the oven to 325°F.

(recipe continues)

EQUIPMENT

Food processor fitted with the steel blade (optional)

9-by-3-inch springform pan or cheesecake pan with removable bottom

Baking sheet

Handheld mixer

Peel off the plastic wrap and place the pan on the baking sheet. Bake for 30 to 35 minutes, checking after 15 to 20 minutes. If the crust has puffed up on the bottom, press it back down carefully with the back of a fork. Continue baking until the crust is golden brown with darker edges. Remove the pan from the oven but leave the oven on. Brush the bottom and sides of the crust carefully with a thin coating of the egg wash. Return the pan to the oven for 2 minutes to set the egg wash. Set the pan on a rack to cool for at least 20 minutes or until you are ready to finish the cake. The crust can be wrapped and kept at room temperature for up to 2 days.

Set the oven temperature to 375°F.

To make the filling, beat the ricotta with the sugar, rice flour, and vanilla with the handheld mixer just until well blended. Beat in the eggs one by one, just until blended. Mix in the candied orange peel or raisins, the almonds, and pine nuts. Scrape the batter into the crust. Bake for 30 minutes. Reduce the oven temperature to 325°F and bake for another 20 to 25 minutes, or until a knife inserted about 2 inches from the edge of the pan comes out clean. The center should still be jiggly. Let cool completely in the pan on a rack before unmolding. Cover and refrigerate for at least 24 hours (48 hours is even better) before serving. Leftovers keep, covered and refrigerated, for another few days.

Patience Required

There is nothing worse than serving a pretty good dessert for a dinner party and finding out 2 days later that it would have been an incredible dessert if you had only waited 2 days to serve it! I've never made a cheesecake that did not improve with at least a full 24 hours, if not 48 hours, of mellowing in the fridge before serving. The difference is astonishing, and this cheesecake is no exception. Make it ahead, way ahead. You won't regret it.

CHESTNUT AND PINE NUT SHORTBREAD

These delicate and not-too-sweet little shortbread cookies are covered in toasted pine nuts—a resonant flavor with chestnuts. Although they may not be authentic, they taste just like Italy to me! Serve them with a little vin santo if you are in Italian mode, or give them as a gift to a sophisticated pal. For a sensational variation, add a cup (100 grams) of chopped walnuts to the dough. **MAKES 45 TO 50 COOKIES**

1½ cups (150 grams) chestnut flour

⅓ cup plus 1 tablespoon (60 grams) white rice flour
—OR—
½ cup plus 1 tablespoon (60 grams) Thai white rice flour

½ cup (100 grams) sugar

Rounded ¼ teaspoon salt

12 tablespoons (1½ sticks/170 grams) unsalted butter, slightly softened and cut into chunks

¼ cup (60 grams) cream cheese

⅓ cup (40 grams) pine nuts

EQUIPMENT

Food processor fitted with the steel blade (optional)

Baking sheets, lined with parchment paper

NOTE: For round cookies, cut the dough with a cookie cutter, then push the scraps back together and cut again; don't worry about cookies becoming tough from reworking the scraps. Or form the dough into a log, wrap, chill, and slice.

To make the dough by hand, put the chestnut and rice flours, sugar, and salt in a large bowl and whisk until thoroughly blended. Add the butter chunks and cream cheese. Use a fork or the back of a large spoon to mash and mix the ingredients together until all are blended into a smooth, soft dough.

To make the dough in a food processor, combine the chestnut and rice flours, sugar, and salt in the food processor and pulse to mix. Add the butter chunks and cream cheese. Pulse until the mixture forms a smooth, soft dough. Scrape the bowl and mix in any stray flour at the bottom of the bowl with your fingers.

Press the dough into an even 9-inch square about ½-inch thick on a sheet of wax paper set on a baking sheet or smaller flat surface. Sprinkle the pine nuts evenly over the dough, then press them gently to embed them in the dough. Cover the dough and refrigerate it for at least 2 hours or, wrapped in plastic, up to 3 days.

Position racks in the upper and lower thirds of the oven. Preheat the oven to 325°F.

Cut the dough into 1¼-inch squares and place them 1 inch apart on the lined pans. Bake for 20 to 25 minutes, until the cookies are golden brown at the edges. Rotate the pans from front to back and top to bottom a little over halfway through the baking. Place the pans on racks or slide the liners from the pans onto racks to cool. Cool the cookies completely before stacking or storing. The cookies may be kept for at least 2 weeks in an airtight container.

CHESTNUT AND WALNUT MERINGUES

Chestnut flour, with or without nuts, makes flavorful meringue cookies, delightful meringue mushrooms (see Variation), and decadent meringues glacées (see page 221). Meringues come together quickly but require lots of time in the oven, so plan accordingly. Meringues also keep well, so you can have cookies, mushrooms, or the makings of a fabulous dessert on hand at all times. **MAKES ABOUT 30 MERINGUE COOKIES**

¼ cup (25 grams) chestnut flour

¾ cup (75 grams) walnut pieces, coarsely chopped

¾ cup (150 grams) sugar

3 large egg whites, at room temperature

¼ teaspoon cream of tartar

EQUIPMENT

Stand mixer with whisk attachment

Baking sheets, lined with parchment paper

Position racks in the upper and lower thirds of the oven and preheat the oven to 200°F.

In a small bowl, mix the chestnut flour and walnuts with ⅓ cup (65 grams) of the sugar.

Combine the egg whites and cream of tartar in the bowl of the stand mixer and beat with the whisk attachment on medium-high speed until the egg whites are creamy white and hold a soft shape when the beaters are lifted. Continue to beat on high speed, adding the remaining sugar a little at a time over 1½ to 2 minutes, until the egg whites are very stiff.

Pour the chestnut flour mixture over the egg whites and fold in with a rubber spatula just until combined.

Drop heaping tablespoons of meringue, or any size and shape you like, 1½ inches apart onto the lined baking sheets.

Bake for 1½ hours. Rotate the pans from top to bottom and from front to back halfway through the baking time to ensure even baking. Remove a "test" cookie and let it cool completely before taking a bite. (Meringues are never crisp when hot.) If the meringue is completely dry and crisp, turn off the heat and let the remaining meringues cool completely in the oven. If the test meringue is soft or chewy or sticks to your teeth, bake for another 15 to 30 minutes before testing again.

To prevent the meringues from becoming moist and sticky, put them in an airtight container as soon as they are cool. They may be stored airtight for weeks.

VARIATION: Chestnut Meringue Mushrooms

Make the meringue as directed, omitting the walnuts. Scrape the meringue into a pastry bag fitted with a plain tip with a ½-inch opening. Pipe pointed "kisses" about 1 inch high to make "stems." Do not worry if the tips bend over or sag. Pipe domes to make mushroom "caps." Sieve a light dusting of unsweetened cocoa powder over the caps and stems and fan them or blow on them vigorously to blur the cocoa and give the mushrooms an authentic look. Bake for 1½ to 2 hours until crisp and completely dry. If not assembling immediately, store the caps and stems airtight as soon as they are cool to prevent them from becoming moist and sticky.

To assemble the mushrooms, place 2 ounces (55 grams) chopped milk or dark chocolate in a small stainless steel bowl set in a skillet of barely simmering water. Immediately turn off heat and stir chocolate until melted and smooth.

Use a sharp knife to cut ¼ to ½ inch off of the tip of each stem—at any angle you like—to create a flat surface. Spread a generous coat of melted chocolate over the flat side of several mushroom caps. Allow the chocolate to set partially before attaching the cut surface of the stems. Repeat until all of the mushrooms are assembled. Set the mushrooms aside until the chocolate has hardened and caps and stems are "glued" together. Store airtight, as soon as possible. (Meringue mushrooms may be made 3 to 4 weeks in advance and stored in an airtight container.) Makes approximately 40 mushrooms with 1¼-inch caps.

CHESTNUT MERINGUES GLACÉES

It is hard to believe that chestnut meringue can raise the bar for the iconic and irresistible combination of sweet and light with crunchy and creamy usually known as meringue glacée. But this new version will knock your socks off—even without embellishments like chocolate sauce or strawberries or luxe chunks of candied chestnuts or candied walnuts. If you keep the meringues on hand, you can pull this spectacular dessert out of your hat with very little notice. Proficiency with a pastry bag will serve you here, but it is not at all necessary. SERVES 12

⅓ cup (35 grams) chestnut flour

1 cup (200 grams) sugar

4 large egg whites, at room temperature

¼ teaspoon cream of tarter

1 quart vanilla or coffee ice cream

2 cups lightly sweetened whipped cream (see page 343), made with a combination of crème fraîche and heavy cream if desired

OPTIONAL GARNISHES

Sliced strawberries

A few candied chestnuts (or drained chestnuts in syrup), chopped

Crushed Walnut Praline Brittle (page 337) or chopped walnuts

Warm chocolate sauce

EQUIPMENT

Stand mixer with whisk attachment

Large pastry bag fitted with a medium to large star tip such as Ateco 9848 or 856 (optional)

Baking sheets, lined with parchment paper

Position racks in the upper and lower thirds of the oven and preheat the oven to 200°F.

In a small bowl, mix the chestnut flour with ⅓ cup (65 grams) of the sugar.

Combine the egg whites and cream of tartar in the bowl of the stand mixer and beat with the whisk attachment on medium-high speed until the egg whites are creamy white and hold a soft shape when the beaters are lifted. Continue to beat on high speed, adding the remaining sugar a little at a time over 1½ to 2 minutes, until the egg whites are very stiff.

Pour the chestnut flour mixture over the egg whites and fold in with a rubber spatula just until combined.

Drop heaping tablespoons of meringue—or use the pastry bag—to make 3-inch rounds or 2-by-4-inch ovals (or any shape you like), 1½ inches apart on the lined sheets.

Bake for 1½ hours. Rotate the pans from top to bottom and from front to back halfway through the baking time to ensure even baking. Remove a "test" cookie and let it cool completely before taking a bite. (Meringues are never crisp when hot.) If the meringue is completely dry and crisp, turn off the heat and let the remaining meringues cool completely in the oven. If the test meringue is soft or chewy or sticks to your teeth, bake for another 15 to 30 minutes before testing again.

To prevent the meringues from becoming moist and sticky, put them in an airtight container as soon as they are cool. They may be stored airtight for weeks.

(recipe continues)

Soften the ice cream in the refrigerator for 15 to 20 minutes, or in the microwave for 10-second bursts. It should be just pliable but not melting.

Spoon 3 to 4 tablespoons of ice cream on the flat bottom of one meringue and spread it level with a spatula, then set a second meringue gently on top to make a sandwich and rotate the top layer gently to seal it rather than pressing it. Immediately set on a baking sheet and place in the freezer. Repeat with the remaining meringues.

Cover the meringues with plastic and freeze until the ice cream is firm enough to serve, or transfer to an airtight container until needed. Meringues glacées keep in the freezer for a few weeks. If the ice cream is frozen rock hard, transfer the meringues to the fridge for 10 to 15 minutes before serving.

To serve, scrape the whipped cream into the pastry bag (if using) and pipe a ruffle of whipped cream on the sides of each meringue glacée, to cover the ice cream. Or simply add a dollop of whipped cream to each serving and garnish, if you like, with berries and/or chestnuts or praline or walnuts; pass the sauce separately.

QUINCE AND ORANGE–FILLED CHESTNUT COOKIES

With a filling of quince paste (available in better supermarkets or specialty stores) mixed with chopped candied orange peel or some grated orange zest, these home-made cookies have a fancy appearance. You can make a date variation by simply substituting chopped moist dates (such as medjools) for the quince paste—with or without the orange peel. **MAKES 20 TO 24 COOKIES**

FOR THE DOUGH

1½ cups (150 grams) chestnut flour

⅓ cup plus 1 tablespoon (60 grams) white rice flour

—OR—

½ cup plus 1 tablespoon (60 grams) Thai white rice flour

½ cup (100 grams) sugar

Scant ½ teaspoon salt

12 tablespoons (1½ sticks/170 grams) unsalted butter, slightly softened, cut into chunks

¼ cup (60 grams) cream cheese

FOR THE FILLING

⅔ cup (180 grams) quince paste

¼ cup (40 grams) finely chopped candied orange peel, or additional quince paste

Grated zest of ½ orange, if not using candied orange peel (optional)

Powdered sugar for dusting

EQUIPMENT

Food processor fitted with the steel blade (optional)

Baking sheets, lined with parchment paper

To make the dough by hand, put the chestnut and rice flours, sugar, and salt in a large bowl and whisk until thoroughly blended. Add the butter chunks and cream cheese. Use a fork or the back of a large spoon to mash and mix the ingredients together until all are blended into a smooth, soft dough.

To make the dough in a food processor, put the chestnut and rice flours, sugar, and salt in the food processor. Pulse to mix. Add the butter chunks and cream cheese. Pulse until the mixture forms a smooth, soft dough. Scrape the bowl and blend in any stray flour at the bottom of the bowl with your fingers.

Press the dough into a ball, wrap it in plastic, and refrigerate it for at least 2 hours, but preferably overnight and up to 3 days. (Or, to save work later, you can fill and shape the cookies immediately and *then* refrigerate them in a covered container for at least 2 hours.)

To make the filling, mash the quince paste with a fork and mix it with the candied orange peel, or orange zest. Shape level teaspoons (8 grams) of the quince paste into little balls (about ¾-inch in diameter) and set them on a plate or piece of wax paper.

Position racks in the upper and lower thirds of the oven. Preheat the oven to 325°F.

Remove the dough from the refrigerator and let it soften for 10 or 15 minutes. Shape level tablespoons of dough (20 to 25 grams) into balls about 1¼ inches in diameter. Make a deep depression in the dough with a knuckle and widen it to form a little bowl. Press a ball of filling into the bowl and ease the dough up around the filling to completely enclose it. Set the cookie, seam side down, on the prepared baking sheets. You can leave the cookies round, or press them into a little beehive shape. Bake for 15 to 20 minutes, until the cookies are slightly golden brown and the bottoms are deep golden. Rotate the pans from front to back and top to bottom a little over halfway through the baking time. Place the pans on racks or slide the liners from the pans onto racks to cool. Sift a little powdered sugar over the cookies. Cool the cookies completely before stacking or storing. The cookies may be kept for at least 2 weeks in an airtight container. Resift with powdered sugar before serving if desired.

The Future of the American Chestnut

The American Chestnut Foundation was founded in 1983 by a group of prominent plant scientists to restore the American chestnut tree to the eastern woodlands on behalf of the environment, wildlife, and society. In 2005, the first potentially blight-resistant chestnuts were harvested. There is no reason why chestnuts and chestnut flours will not flourish once again, less as a staple food than a flavorful and nutritious ingredient. In the meantime, there are several domestic sources of chestnut flour, of which Allen Creek Farm is the best for these recipes (their flour is not at all smoky; see Resources, page 351).

CHESTNUT PUDDING

Chestnut pudding does not require the usual cornstarch or rice flour thickener, as the flour itself provides both the flavor and the thickener. The pudding has a slightly nubbly, almost porridgey texture, like semolina pudding or rice pudding with bits of chestnut instead of rice grains. It's a homey dessert. Top with crunchy toasted chopped walnuts (or Walnut Praline Brittle, page 337) and/or a spoonful of apple or pear butter. **SERVES 6 TO 8**

⅔ cup (70 grams) chestnut flour

¼ cup plus 2 tablespoons (75 grams) sugar

Generous pinch of salt

1 cup whole milk

3 cups half-and-half

EQUIPMENT

Six 6-ounce custard cups or ramekins or 8 smaller cups

In a heavy medium saucepan, whisk the chestnut flour, sugar, and salt with just enough of the milk to make a loose paste. Whisk in the remaining milk and the half-and-half. Cook over medium heat, stirring constantly with a silicone spatula or a wooden spoon, scraping the bottom, sides, and corners of the pan, until the mixture thickens and begins to bubble at the edges. Set a timer for 2½ minutes and continue to cook and stir, adjusting the heat so that the mixture bubbles readily but not furiously.

Divide the pudding among the cups. Serve warm, at room temperature, or chilled. The pudding can be refrigerated, covered, for up to 3 days.

CHESTNUT PRALINE GELATO

Sicilian gelato is one of my favorite types of ice cream. It's rustic and both extra flavorful and extra refreshing because it's not super rich. And, since it has no eggs, it's wildly easy to make. Chestnut flour flavors and thickens the ice cream base, in lieu of the usual cornstarch. Crushed caramelized nuts add flavor and crunch and completely fool your palate by making the otherwise slightly textured (from the grainy chestnut flour) ice cream seem completely smooth. **MAKES ABOUT 1 QUART**

Scant ½ cup (50 grams) chestnut flour

3 tablespoons (35 grams) sugar

¼ teaspoon salt

3 cups half-and-half

½ teaspoon pure vanilla extract

1 cup chopped praline made with walnuts or toasted skinned hazelnuts (see page 309)

EQUIPMENT

Ice cream maker

Put the chestnut flour, sugar, and salt in a medium saucepan. Whisk in enough of the half-and-half to make a smooth paste. Whisk in the remaining half-and-half and the vanilla. Cook over medium heat, stirring with the whisk and sweeping the bottom, sides, and corners of the pot to prevent scorching, until the mixture comes to a simmer. Continue to cook and stir, adjusting the heat to maintain a lively simmer, for 3 minutes to cook the flour completely. Scrape the mixture into a bowl and let cool at room temperature or in the refrigerator.

Freeze the mixture according to the instructions with your ice cream maker, adding the praline at the end. Store the gelato in the freezer until needed. If the gelato is rock hard, let it soften in the refrigerator for 15 minutes or more before serving or in the microwave on Defrost for a few seconds at a time until scoopable. The gelato is best within 2 or 3 days.

CHAPTER SIX

TEFF FLOUR

AMERICANS who have not been to an Ethiopian restaurant, are not gluten intolerant, or do not have an avid interest in healthy grains may not know teff. Aptly considered an "ancient grain," it is a species of lovegrass native to the East African highlands of Ethiopia, where it represents one-quarter of the country's cereal production and supplies Ethiopians with an estimated two-thirds of their dietary protein.

Whole teff grains are cooked and eaten like porridge. The flour milled from the grain is used to make the daily bread of Ethiopia, the spongy crepe-like sourdough flat bread called *injera* that is served with and used as a utensil at most every meal. Of all the flours used in this book, teff has the closest tie to one geographical area—and to one specific recipe, which, though delicious, is blissfully ignored in this chapter.

The teff seed is 150 times smaller than a grain of wheat and consists mostly of germ and bran. It would be difficult to separate the parts of this almost microscopic seed, so teff flour is ground from the whole grain. It is loaded with calcium, protein, iron, vitamin C, and fiber as well as other nutrients. But my interest in teff flour transcends its stellar nutritional credentials and ancient pedigree.

Teff flour is a very fine, free-flowing speckled flour with a slatey gray-brown color—slightly more reddish than buckwheat. Stick your nose into a freshly opened bag and take in a gentle malted, toasted aroma suggestive of Thai iced tea or Chinese milk tea (a combination of black tea, spices, and cooked milk). The scent just hints at the flavors to come.

My first teff sponge cake, right out of the oven, had an aroma of sweet whole wheat—whole wheat with none of its usual bitterness—and hints of natural cocoa and hazelnut. It also tasted sweeter than the same cake made with all-purpose (wheat) flour simply from the inherent sweetness of the teff.

Teff flour may produce a fine crystalline texture in some lighter baked goods; cocoa powder softens this effect in some recipes and ground nuts balance it in others. Teff can turn gummy in other recipes (for example, butter cake) unless there are other textural elements to offset the effect.

Pairing teff with cocoa seems natural: chocolate génoise made with teff flour may be the best chocolate génoise you will ever taste—and perfect for all kinds of fancy layer cakes. It also makes lighter *and* more flavorful layers for that otherwise richer-than-sin (and rather heavy) retro classic, German Chocolate Cake (page 237). Teff adds a sweet whole wheat note and beautiful color to a dense fruit and nut cake, buttery short crusts for tarts, and dainty flavorful crepes to fill. Teamed up with flax, pumpkin, caraway, and cumin seeds, teff makes stellar crackers. Use teff to make killer brownies, too.

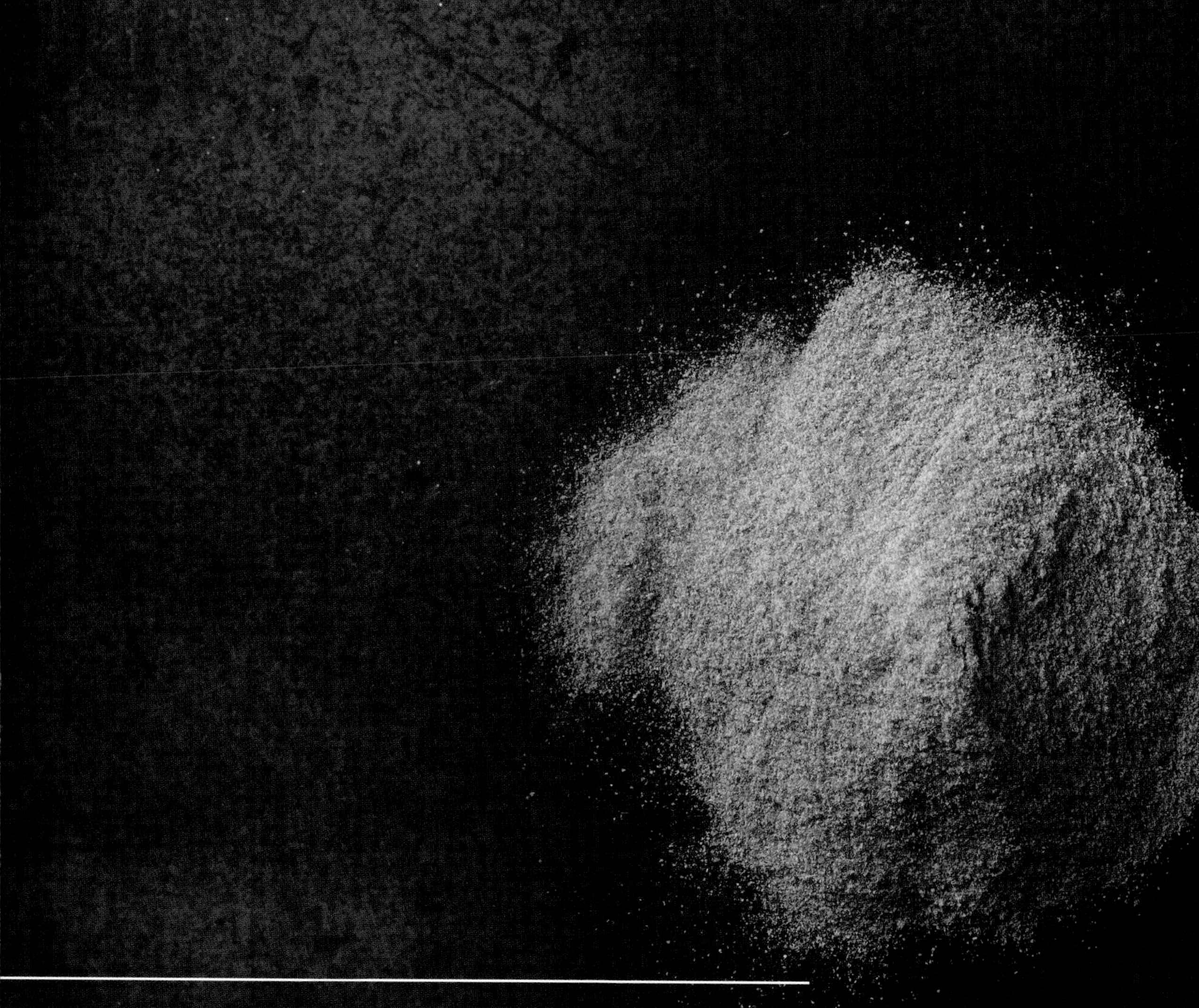

FLAVOR AFFINITIES FOR TEFF FLOUR

Nuts, chocolate, dark fruit, aromatic seeds

WHERE TO BUY AND HOW TO STORE

Teff is available in better supermarkets in the specialty flour section of the baking aisle, or by mail order from the Teff Company or Bob's Red Mill (see Resources, page 351). Teff flour is a whole grain and should be stored in an airtight container, away from heat and light, for 2 to 3 months at room temperature, or 6 months in the refrigerator, and up to 12 months in the freezer.

THE NEW CHOCOLATE GÉNOISE

A combination of teff flour and natural cocoa powder produces an outstanding génoise. If you are a seasoned home baker, or a professional with a repertoire of European desserts, you will find this cake to be a perfect swap for the usual chocolate génoise in a multitude of desserts. But it's not just good enough to treat in the classic way—by brushing with liqueurs and filling and frosting with buttercreams and mousses—it's also tempting enough to eat by itself or with a little whipped cream and some berries. The layer cakes that follow are just a sample of what you can do with this cake. **SERVES 10 TO 12**

---

3 tablespoons (45 grams) Clarified Butter (page 335) or ghee (see page 38)

¼ cup plus 1 tablespoon (30 grams) unsweetened natural cocoa powder

½ cup (65 grams) teff flour

4 large eggs

⅔ cup (130 grams) sugar

⅛ teaspoon salt

EQUIPMENT

8-by-3-inch springform pan or cheesecake pan with removable bottom

Fine strainer or tea strainer

Stand mixer with whisk attachment

Sifter or medium-mesh strainer

Position a rack in the lower third of the oven and preheat the oven to 350°F. Line the bottom of the pan with parchment paper, but do not grease the sides of the pan.

Put the clarified butter in a small pot or microwavable container ready to reheat when needed, and have a 4- to 5-cup bowl ready to pour it into as well—the bowl must be big enough to allow you to fold some batter into the butter later.

Press the cocoa through a fine strainer into a medium bowl. Add the flour and whisk to blend.

Combine the eggs, sugar, and salt in the bowl of the stand mixer and beat with the whisk attachment on high speed for 4 to 5 minutes. The mixture should be light colored and tripled in volume, and you should see well-defined tracks as the whisk spins; when the whisk is lifted, the mixture should fall in a thick, fluffy rope that dissolves slowly on the surface of the batter.

Just before the eggs are ready, heat the clarified butter until very hot and pour it into the reserved bowl.

Remove the bowl from the mixer. Sift one-third of the flour over the eggs. Fold with a large rubber spatula until the flour is almost blended into the batter. Repeat with half of the remaining flour. Repeat with the rest of the flour. Scrape about one-quarter of the batter into the hot butter. Fold until the butter is completely blended into the batter. Scrape the buttery batter over the remaining batter and fold just until blended. Scrape the batter into the pan.

Bake for 30 to 35 minutes, until the cake has puffed up and then settled level but hasn't pulled away from the sides of the pan and a toothpick inserted in the center comes out clean. Set the pan on a rack. While the cake is still hot, run a small spatula around the inside of the pan, pressing against the sides of the pan to avoid tearing the cake.

At your convenience (the cake can be warm or completely cool), invert the pan onto the rack to remove the cake and peel off the parchment liner. Turn the cake right side up to finish cooling. The cake should be completely cool before filling, frosting, or storing. The cake may be wrapped airtight and stored at room temperature for 2 days, or frozen for up to 3 months.

VARIATION: Brown Butter Chocolate Génoise

Have a small stainless steel bowl near the stove. Melt 4 tablespoons (½ stick/55 grams) unsalted butter in a medium saucepan until it bubbles. Continue to cook, whisking gently, until the butter is golden brown and the milk particles suspended in it are reddish brown. Immediately pour the brown butter into the stainless bowl to stop the browning. Use all of the brown butter in place of the clarified butter in the recipe.

Crib Sheet for Making 1½ Recipes of The New Chocolate Génoise

The layer cake recipes that follow call for 1½ recipes of chocolate génoise batter baked in three 8-inch pans to make three separate layers. Here are the amounts of each ingredient, so you don't have to do the math.

4½ tablespoons (70 grams) Clarified Butter (page 335) or ghee (see page 38)

Scant ½ cup (45 grams) unsweetened natural cocoa powder

¾ cup (100 grams) teff flour

6 large eggs

1 cup (200 grams) sugar

Scant ¼ teaspoon salt

The beating time is 6 or 7 minutes. The baking time is 15 to 20 minutes (until cakes begin to pull away from the sides of the pans) in three 8-inch pans, ungreased but lined with parchment paper. The cakes are cooled upside down on wax or parchment paper.

TRIPLE CHOCOLATE LAYER CAKE

This is a reimagined and updated version of one of my favorite all-chocolate layer cakes. I created the original with the extraordinary baker and cookbook author Flo Braker, to honor Chuck Williams—founder of Williams-Sonoma stores—on his eightieth birthday. Julia Child was present at the party and stood up spontaneously to toast the cake (and Chuck, of course)!

Chocolate stars in three different ways here: a tender cocoa sponge cake, a light but dreamy whipped chocolate filling, and an intense, deeply bittersweet chocolate glaze. This is a perfect sophisticated birthday cake for an adult, though I've never seen a kid walk away from it. **SERVES 12 TO 16**

Batter for 1½ recipes of The New Chocolate Génoise (page 232) or Brown Butter Chocolate Génoise (page 233)

Whipped Ganache Filling (recipe follows), chilled but not whipped

Sarah Bernhardt Chocolate Glaze (page 347), cooled to the consistency of frosting

EQUIPMENT

Three 8-inch round cake pans, 1½ to 2 inches deep

Instant-read thermometer

NOTE: For an extra moist and decadent dessert, you can brush the layers with syrup before filling them: Combine 3 tablespoons rum, brandy, bourbon, or eau de vie with 3 tablespoons sugar and 2 tablespoons water. Stir and let stand for about 15 minutes to dissolve the sugar.

Position racks in the upper and lower thirds of the oven and preheat the oven to 350°F. Line the bottoms of the cake pans with parchment paper and leave the sides ungreased.

Make the batter as directed, but beat the eggs and sugar a minute or two longer, and divide it evenly among the pans. Bake for 15 to 20 minutes, until the cakes begin to pull away from the sides of the pans. Rotate the pans from back to front and upper to lower about halfway through the baking time.

When the cakes are done, run a small spatula around the inside of each pan, pressing against the sides of the pan to avoid tearing the cake. Invert each cake onto a piece of wax or parchment paper on a baking sheet. Remove the pan and peel off the parchment liner. Leave the cakes to cool upside down on the wax paper. When cool, turn each cake over and peel off the wax paper—the skin on the surface of the cake will come off with the wax paper, saving you the extra step of trimming it off later.

Set one cake layer, right side up, on a cardboard cake circle or the bottom of an 8-inch springform pan. Whip the ganache as directed in the recipe—not too stiff—and (working quickly before the ganache stiffens) spread half of it evenly over the first cake layer. Top with a second cake layer and spread the remaining ganache over it. Top with the third layer, bottom side up. Press gently on the cake to compact and level it. Chill the cake for at least 2 hours.

To finish the cake, spread a thin layer of the cooled glaze over the top and sides of the cake, just to smooth the surfaces, glue on any crumbs, and fill the cracks between the layers—this is called the crumb coat. Refrigerate the cake while rewarming the glaze.

Set the bowl with the remaining glaze in a wide skillet of not-even-simmering water; stir gently with a silicone spatula until the glaze is fluid and shiny. Check the temperature of the glaze with an instant-read thermometer; if it exceeds 90°F, let the glaze cool, stirring gently, until it is 88°F to 90°F.

Center the cake on a lazy Susan or turntable if you have one, or else on a baking sheet. Have ready a clean, dry metal icing spatula. Scrape all of the glaze onto the center of the cake. Working quickly, use just three or four strokes to spread the glaze over the top of the cake so that it runs down to coat the sides. If there are any bare spots left uncoated, use the spatula to scoop up excess glaze and touch it to the bare spots, without spreading.

Refrigerate the cake immediately to set the glaze, and then put the cake in a covered container or under a cake dome.

The cake keeps in the refrigerator, covered, for up to 3 days. Remove 30 to 60 minutes before serving.

(recipe continues)

A Better Way to Remove Skins from Génoise Layers

The skin that forms on the top of génoise or sponge layers is moist and tasty, but it can get in the way: the skin may peel up when you try to spread a filling over it, or it may simply prevent the filling from attaching to the cake beneath it. For these reasons, pastry chefs often trim the skin before making layered desserts. Trimming the skin with a knife is messy and annoying—one of my least favorite tasks in the pastry kitchen. Here's the better way: When cake layers come out of the oven, run a spatula around the sides to detach them from the pans and then invert them onto sheets of wax or parchment paper on a baking sheet. Remove the pans and peel off the parchment liners to let the cakes breathe, but leave them to cool (and stick) on the wax paper. When the cake layers are cool, turn them right side up and peel off the wax paper—the cake skins will come off too. Nothing could be cleaner or easier.

Whipped Ganache Filling

Prepare and chill the ganache for at least 6 hours or up to 4 days in advance, but don't whip it until you are ready to use it. **MAKES ABOUT 3½ CUPS (AFTER WHIPPING)**

8 ounces (225 grams) 54% to 60% chocolate, chopped medium-fine

2 cups heavy cream

EQUIPMENT

Stand mixer with paddle attachment, or handheld mixer

Place the chopped chocolate in a medium bowl. Heat the cream in a medium heavy saucepan until it comes to a gentle boil. Immediately pour the cream over the chocolate and wait a minute or two before stirring the chocolate into the cream. Let the mixture stand for 15 to 20 minutes to be certain all of the chocolate is melted. Then stir again—the ganache should be perfectly smooth, without any unmelted chocolate specks. Let the ganache cool, then cover and refrigerate for at least 6 hours or up to 4 days; the ganache must be perfectly cold before you whip it. When you are ready to use the ganache (and not a moment before), whip it using a stand mixer with the paddle attachment or a handheld mixer until it is stiff enough to hold a nice soft shape and is easily spreadable. (Don't whip it stiff or it will be grainy and difficult to spread.)

GERMAN CHOCOLATE CAKE

A retro favorite gets a tune-up here with a lighter but more flavorful cake and better-than-ever filling—improved by simply toasting the pecans and coconut—that you'll want to eat with a spoon. **SERVES 16**

Batter for 1½ recipes of The New Chocolate Génoise (page 232)

4 large egg yolks

One 12-ounce can evaporated milk or 1½ cups fresh half-and-half

1½ cups (300 grams) sugar

¾ teaspoon salt

2 teaspoons pure vanilla extract

12 tablespoons (1½ sticks/170 grams) unsalted butter, cut into several chunks

7 ounces (about 2⅔ cups/200 grams) sweetened shredded dried coconut, toasted

1½ cups (150 grams) pecan halves, toasted, then chopped

EQUIPMENT

Three 8-inch round cake pans, 1½ to 2 inches deep

Position racks in the upper and lower thirds of the oven and preheat the oven to 350°F. Line the bottoms of the cake pans with parchment paper and leave the sides ungreased.

Make the batter as directed, but beat the eggs and sugar a minute or so longer, and divide it evenly among the pans. Bake for 15 to 20 minutes, until the cakes begin to pull away from the sides of the pans. Rotate the pans from back to front and upper to lower about halfway through the baking time.

When the cakes are done, run a small spatula around the inside of each pan, pressing against the sides of the pan to avoid tearing the cake. Invert each cake onto a piece of wax or parchment paper on a baking sheet. Remove the pan and peel off the parchment liner. Leave the cakes to cool upside down on the wax paper. When cool, turn each cake over and peel off the wax paper—the skin on the surface of the cake will come off with the wax paper, saving you the extra step of trimming it off later (see box, page 235).

To make the filling, in a heavy nonreactive saucepan, whisk the egg yolks with the milk, sugar, salt, and vanilla. Add the butter. Cook over medium heat, stirring constantly with a silicone spatula, scraping the bottom, sides, and corners of the pot. When the mixture starts to boil, adjust the heat so that it boils actively but not furiously. Cook until the mixture is golden brown and thickened, about 12 minutes total cooking time. Take the pan off the heat and stir in the coconut and pecans. Cool completely before using.

To assemble the cake, set one cake layer on a serving platter. Spread with one-third of the filling. Top with a second cake layer. Spread with half of the remaining filling. Put the third cake layer on top and cover it with the remaining filling. Leave the sides exposed. Serve at room temperature.

QUEEN OF THE NILE

How could I resist adding teff flour, the indigenous grain of Ethiopia, to the chocolate cake that helped launch my career? The iconic torte Reine de Saba—Queen of Sheba—is named for the eponymous Ethiopian queen. This new torte is rich and delicious, soft textured and very chocolaty. It is everything it should be, and it comes together quickly with a handheld mixer in the same bowl used for melting the chocolate. You may trade pecans or toasted hazelnuts for the almonds to vary the flavor. Serve the torte simply, dusted with a little powdered sugar, or plate it with a dollop of whipped cream. **SERVES 10 TO 12**

½ cup (2½ ounces/70 grams) whole almonds (with skins or blanched),, or ⅔ cup (2½ ounces/70 grams) almond flour/meal

¼ cup (35 grams) teff flour

6 ounces (170 grams) 70% chocolate, coarsely chopped

¾ cup (150 grams) sugar

10 tablespoons (1¼ sticks/140 grams) unsalted butter, softened but not too squishy, cut into chunks

⅛ teaspoon salt

4 large cold eggs

EQUIPMENT

8-by-3-inch springform pan or cheesecake pan with removable bottom

Food processor fitted with the steel blade

Handheld mixer

NOTE: The trick to getting a fluffy aerated batter is to be sure that the chocolate is not too warm and the butter not too soft, and the eggs must be cold!

Position a rack in the lower third of the oven and preheat the oven to 375°F.

Grease the bottom and sides of the pan with vegetable oil spray or butter.

Pulse the almonds and teff flour in a food processor until finely ground. Or simply mix the almond flour with the teff. Set aside.

Put the chocolate in a large stainless steel bowl, set it in a wide skillet of barely simmering water, and stir occasionally until the chocolate is nearly melted. Remove the bowl from the water bath and stir the chocolate until completely melted and smooth. Let cool to lukewarm (if necessary). Add the almond-teff mixture, sugar, butter chunks, and salt to the chocolate and beat with the handheld mixer on medium speed until the ingredients are well blended and the mixture thickens and lightens slightly in color. Beat in the eggs, one by one. Continue beating on high speed for a minute or two, or until the batter is fluffy and lighter in color; it should resemble fluffy chocolate frosting (see Note).

Scrape the batter into the prepared pan and spread it evenly. Bake for 30 to 35 minutes, until a toothpick inserted in the center of the cake comes out with a few moist crumbs. Set the pan on a rack to cool. Slide a thin knife or a small metal spatula around the inside of the pan to loosen the cake and allow the thin crust on top to sink (slightly) as the cake cools. Let cool completely. Remove the pan sides and transfer the cake to a serving platter. The cake can be kept at room temperature, covered or under a cake dome, for up to 3 days, or frozen, well wrapped, for up to 3 months; bring to room temperature before serving.

CHOCOLATE RASPBERRY CELEBRATION CAKE

The moment I tasted The New Chocolate Génoise, I knew that I had to update the cake that I made on TV with Julia Child on *Baking at Julia's* many years ago. A great cake recipe can always be made better. Génoise made with teff and cocoa does not require as much soaking syrup as the old standard, and I have even made the cake layers a little thicker, because they are so delicious. Meanwhile, over the years, people have asked how to make the cake without using any alcohol at all. Simply replace the spirits or liqueur with ¼ cup strained raspberry puree and reduce the water to 1 tablespoon and the sugar to 4 teaspoons. **SERVES 12 TO 16**

Batter for 1½ recipes of The New Chocolate Génoise (page 232)

3 tablespoons eau de vie de framboise or raspberry vodka (clear high proof raspberry spirits, not sweet pink liqueur), or ¼ cup plus 2 tablespoons sweet raspberry liqueur, such as Chambord

3 tablespoons sugar (if using eau de vie or vodka)

2 tablespoons water (if using eau de vie or vodka)

2½ cups chilled crème fraîche

1½ teaspoons pure vanilla extract

Sugar to taste

4 ounces 60% to 62% chocolate, chopped (see Note on using other chocolates)

3 tablespoons water

2 cartons (10 ounces/280 grams) fresh raspberries

Sarah Bernhardt Chocolate Glaze (page 347), cooled to the consistency of soft frosting

To make the cakes, position racks in the upper and lower thirds of the oven and preheat the oven to 350°F. Line the bottoms of the cake pans with parchment paper and leave the sides ungreased.

Make the batter as directed, but beat the eggs and sugar a minute or two longer, and divide it evenly among the three pans. Bake for 15 to 20 minutes, until the cakes begin to pull away from the sides of the pans. Rotate the pans from back to front and upper to lower about halfway through the baking time.

When the cakes are done, run a small spatula around the inside of each pan, pressing against the sides of the pan to avoid tearing the cake. Invert each cake onto a piece of wax or parchment paper on a baking sheet. Remove the pan and peel off the parchment liner. Leave the cakes to cool upside down on the wax paper. When cool, turn each cake over and peel off the wax paper—the skin on the surface of the cake will come off with the wax paper, saving you the extra step of trimming it off later (see box, page 235).

To make the framboise syrup, if using eau de vie or raspberry vodka, combine it with the sugar and water in a small cup or jar, stir, cover, and set aside for at least 15 minutes to dissolve the sugar. If using sweet raspberry liqueur, simply pour it into a cup or jar; do not add the sugar or water.

To assemble the cake, beat the chilled crème fraîche with the vanilla and sugar to taste until very thick but not stiff. Refrigerate until needed.

Fit one cake layer, right side up, into the bottom of the springform pan. Use a pastry brush to moisten the layer with about one-third of the framboise syrup.

EQUIPMENT

Three 8-inch round cake pans, 1½ to 2 inches deep

8-by-3-inch springform pan or cheesecake pan with removable bottom

Pastry brush

Instant-read thermometer

Combine the chocolate and water in a small stainless steel bowl. Set the bowl in a wide skillet of barely simmering water and stir until the chocolate is barely melted. Remove from the water and stir until the chocolate is completely melted and smooth. Wait until the chocolate is barely lukewarm. Measure out ¾ cup of the whipped crème fraîche. Working quickly, fold one-third of that amount—about ¼ cup—into the tepid chocolate mixture and then fold in the remaining ½ cup.

Immediately (before the mixture stiffens) scrape all of the chocolate crème fraîche on top of the moistened cake layer and spread it evenly.

Moisten the second génoise layer—with about half the amount of the framboise syrup used on the bottom layer—and fit it moist side down over the chocolate crème fraîche. Press in place. Moisten the top with about the same amount of framboise syrup. Distribute the raspberries in one layer over the cake, leaving a little space around each one.

Whip the remaining crème fraîche to make it a bit stiffer. Scoop it over the berries. Spread and press the crème fraîche over and between the berries and against the sides of the pan.

Moisten the third cake layer with half of the remaining framboise syrup and fit it moist side down into the pan. Press in place. Moisten the top with the remaining syrup. Cover the pan with plastic wrap and refrigerate the assembled cake for at least 2 hours or up to 24 hours.

To crumb coat and glaze the cake, remove the sides of the cake pan. Spread a thin layer of the cooled glaze over the top and sides of the cake, just to smooth the surfaces, glue on any crumbs, and fill the cracks between the layers. Refrigerate the crumb-coated cake while reheating the glaze.

Remelt the remaining glaze very gently by setting the bowl in a wide skillet of not-even-simmering water; stir gently with a silicone spatula until the glaze is fluid and shiny. Check the temperature of the glaze with an instant-read thermometer: if it exceeds 90°F, let it cool, stirring gently, until it is 88°F to 90°F.

Center the cake on a lazy Susan or turntable, if you have one, or else on a baking sheet. Have ready a clean, dry metal icing spatula.

(recipe continues)

Scrape all the glaze onto the center of the cake. Working quickly, use just three or four strokes, rotating the turntable, if using, to spread the glaze over the top of the cake so that it runs down to coat the sides. If there are any bare spots left uncoated with glaze, use the spatula to scoop up excess glaze and touch it to the bare spots, without spreading.

Refrigerate the cake immediately to set the glaze. The cake keeps for a day or two, covered, in the refrigerator. Remove the cake from the refrigerator 30 to 60 minutes before serving.

NOTE: To use 55% chocolate, increase the amount of chocolate to 5 ounces/140 grams. To use 70% chocolate, use 3½ ounces/100 grams of chocolate and increase the water to 3½ tablespoons with 2 to 3 teaspoons of sugar dissolved in it.

DATE-NUT CAKE WITH APRICOTS AND TEFF

Rather than just helping to glue the abundant fruits and nuts together, teff flour in the batter adds its own flavor mojo and a rich mahogany color to this versatile mosaic loaf. You can fiddle with the recipe by switching fruits and/or tweaking the proportions. You can also make it a little sweeter or less sweet. This treasured recipe can be your healthy snack, teatime treat, or dessert. Try it with an aged Oloroso sherry. One of my favorite Berkeley restaurants sprinkles the top of its date cake (also my recipe!) with a little coarsely ground black pepper just before baking; slices are served with a cheese course. **MAKES 1 LOAF / PHOTO ON PAGE 176**

½ cup (70 grams) teff flour

⅛ teaspoon baking soda

⅛ teaspoon baking powder

¼ teaspoon salt

3 to 4 tablespoons (35 to 50 grams) packed light or dark brown sugar

1 cup (175 grams) dates, pitted and cut into quarters

½ cup (75 grams) lightly packed dried apricot halves, or a combination of dried apricots and dried pluots, cut in half

1¾ cups (175 grams) walnut pieces

2 large eggs

1¼ teaspoons pure vanilla extract

EQUIPMENT

8-by-4-inch (4-cup) loaf pan, bottom and all four sides lined with parchment paper

Position a rack in the lower third of the oven and preheat the oven to 300°F.

In a large bowl, whisk the flour with the baking soda, baking powder, and salt. Add the brown sugar, dates, apricots, and nuts and mix thoroughly with your fingers, separating any sticky fruit pieces from one another. Set aside.

Whisk the eggs and vanilla in a medium bowl until lightened in color. Scrape the eggs into the large bowl and mix with a spatula until all of the fruit and nut pieces are coated with batter. Scrape into the prepared pan.

Bake for 60 to 80 minutes, until the top is mahogany brown and feels firm and crusted. Cool in the pan on a rack. Lift the ends of the paper liner and transfer the cake to a cutting board. Use a sharp knife to cut very thin slices. The cake may be stored, airtight, for at least 1 week at room temperature, or longer in the refrigerator.

BLACK CHERRY CHOCOLATE LINZER TORTE

A chocolate linzer torte is a difficult dessert to perfect. The chocolate can either overwhelm this classic rich sweet or simply not play nicely with the other flavors. It took teff flour—which tastes a little like cocoa anyway—and just a little natural cocoa powder to get the balance right. The teff is complemented by the spices, the hazelnuts, and some cherry preserves instead of the classic black currant or raspberry.

To make things easy, there is no formal lattice on top. It's been replaced with a random scattering of reserved dough either cut into matchsticks or coarsely grated on the largest holes of a handheld grater. Linzer torte keeps well for at least a week and is delicious all by itself or with vanilla ice cream or whipped cream. **SERVES 12**

¾ cup (110 grams) raw hazelnuts

½ cup (80 grams) white rice flour
—OR—
¾ cup (80 grams) Thai white rice flour

⅓ cup (40 grams) teff flour

1 tablespoon plus 1 teaspoon (8 grams) unsweetened natural cocoa powder

¾ cup (150 grams) granulated sugar

¼ teaspoon salt

1¼ teaspoons ground cinnamon

¼ teaspoon ground cloves

11 tablespoons (155 grams) unsalted butter, slightly softened and cut into chunks

1 large egg white

⅔ cup (210 grams) cherry preserves

1 to 2 tablespoons powdered sugar for dusting (optional)

EQUIPMENT

Food processor fitted with the steel blade

9½-inch fluted tart pan with removable bottom or 9-inch round cake pan

Baking sheet

In the bowl of the food processor, combine the hazelnuts, rice and teff flours, cocoa, granulated sugar, salt, cinnamon, and cloves. Pulse until the hazelnuts are finely ground. Add the butter and egg white. Pulse just until blended.

Measure ¼ cup of the dough (65 grams) and shape it into a cube. Wrap and refrigerate it until needed.

Meanwhile, spray the sides of the tart or cake pan with vegetable oil spray and line the bottom with a round of parchment. Press the remaining dough evenly over the bottom but not up the sides of the pan. Cover and refrigerate for at least an hour.

Position a rack in the lower third of the oven. Preheat the oven to 350°F.

Spread the preserves evenly over the dough, leaving a scant ½-inch border all around. Cut the chilled reserved cube of dough into thin slices and then matchsticks. Scatter the matchsticks randomly over the preserves; don't worry if some of them break. (Or, using the largest holes of a handheld grater, grate the dough over the jam to make a scattered pattern.)

Set the pan on the baking sheet. Bake until golden brown, 30 to 35 minutes. Tent the torte loosely with foil and continue baking for 10 to 15 minutes, until the crust is a deep golden brown. If the torte puffs up during baking, rap the pan sharply on the oven rack to settle it. Set the pan on a rack to cool.

(recipe continues)

After 10 minutes, loosen the rim of the tart pan all around by pushing the pan bottom up gently (you will need pot holders to do this). Or, if using a solid cake pan, run a slim knife or spatula around the insides to detach the cake. When the torte is completely cool, invert it onto a plate. Remove the pan and the parchment liner and turn the torte right side up. The torte is most delicious served within 3 or 4 days but still remarkably good after a week. Store and serve at room temperature, sprinkled with powdered sugar, if desired.

Teff Is Here to Stay

Surprisingly, considering how few Americans have heard of it even now, teff has been cultivated in the United States since the 1980s because Wayne Carlson, a former Peace Corps volunteer, fell in love with Ethiopian culture and food and decided to bring the crop home to Idaho. As a result of his pioneering work with (and steadfast cajoling of) farmers, teff is now a popular specialty crop in several Western states, with a growing market in Ethiopian communities here and abroad, as well as those searching for gluten-free and/or healthy grains. Meanwhile, its high yield and good quality make teff an exceptional alternate forage crop in rotation with legumes such as alfalfa. By 1996, the United States National Research Council had identified teff as having "potential to improve nutrition, boost food security, foster rural development, and support sustainable land care." All things considered, teff flour is likely to become better known to all of us in the coming years. Carlson's Teff Company (see Resources, page 351) remains a good source for ordering teff and teff flour online.

MOCHA CREAM TART WITH CHOCOLATE CRUST

Cocoa powder makes a beautifully creamy rich mocha custard filling for this easy tart in a chocolate shortbread crust. Serve the tart slightly warm, cool, or cold, sprinkled with cocoa powder if you like. **SERVES 8 TO 10**

FOR THE CRUST

½ cup plus 1 tablespoon (75 grams) teff flour

3 tablespoons (30 grams) white rice flour
—OR—
Scant ⅓ cup (30 grams) Thai white rice flour

¼ cup (50 grams) sugar

3 tablespoons (18 grams) unsweetened cocoa powder (natural or Dutch-process)

¼ teaspoon salt

6 tablespoons (85 grams) unsalted butter, slightly softened and cut into chunks

2 tablespoons (30 grams) cream cheese

1½ teaspoons water

½ teaspoon pure vanilla extract

1 egg yolk mixed with a pinch of salt and ½ teaspoon water, for the egg wash

FOR THE FILLING

4 tablespoons (½ stick/55 grams) unsalted butter, cut into chunks

⅔ cup (130 grams) sugar

¼ cup plus 2 tablespoons (35 grams) unsweetened cocoa powder (natural or Dutch-process)

1½ cups heavy cream

1 teaspoon instant espresso powder or 1¼ teaspoons instant coffee crystals

¾ teaspoon pure vanilla extract

1 large egg plus 1 egg yolk

Grease the pan lightly with vegetable oil spray or butter.

To make the crust by hand, put the teff and rice flours, sugar, cocoa, and salt in a medium bowl and whisk until thoroughly blended. Add the butter chunks, cream cheese, water, and vanilla. Use a fork or the back of a large spoon to mash and mix the ingredients together until all are blended into a smooth, soft dough.

To make the crust in a food processor, put the teff and rice flours, sugar, cocoa, and salt in the food processor. Pulse to blend. Add the butter chunks, cream cheese, water, and vanilla. Pulse until the mixture forms a smooth, soft dough. Scrape the bowl and blend in any stray flour at the bottom of the bowl with your fingers.

Transfer the dough to the tart pan.

The dough may seem much softer than other tart doughs. Use the heel of your hand and then your fingers and/or a small offset spatula to spread the dough all over the bottom of the pan. Press it squarely into the corners of the pan with the side of your index finger to prevent extra thickness at the bottom edges, and press it as evenly as possible up the sides of the pan, squaring it off along the top edge. Have patience; there is just enough dough (although you may not think so at first). If there is too much dough in one place (or hiding in the corners of the pan), pinch or scrape it off and move it elsewhere. Spread or smear it smooth with the spatula. Here's a final trick for a perfectly even crust: Press a sheet of plastic wrap against the bottom and up the sides of the pan and lay a paper towel on top. Set a straight-sided, flat-bottomed cup on the towel; press and slide the cup all over the bottom and around the sides to smooth and even the surface. Leave the plastic wrap

EQUIPMENT

9½-inch fluted tart pan with a removable bottom

Food processor fitted with the steel blade (optional)

Rimmed baking sheet

in place. Refrigerate the pan for at least 2 hours, but preferably overnight and up to 2 days.

Position a rack in the lower third of the oven and preheat the oven to 325°F.

Place the pan on the baking sheet and peel off the plastic wrap. Bake for 35 to 40 minutes, checking after 15 to 20 minutes. If the crust has puffed up on the bottom, press it back down carefully with the back of a fork if necessary. Continue baking until the crust is a slightly deeper shade of brown and has pulled away from the edges of the pan all around.

Meanwhile, start the filling: Place the butter, sugar, cocoa powder, and cream in a medium saucepan and cook over medium heat, stirring, until the mixture is blended and smooth and begins to simmer around the edges. Remove from the heat and stir in the espresso powder and vanilla. Set aside.

When the crust is done, remove it from the oven but leave the oven on. Brush the bottom and sides of the crust gently with a thin coat of the egg wash (discard any remaining egg wash or save for another use). Return the crust to the oven for 2 minutes to set the egg wash.

When the egg wash is set, whisk the egg and the egg yolk thoroughly into the hot filling. Pour the filling into the hot crust and turn the oven down to 300°F. Leave the tart in the oven until the filling quivers like tender Jell-O (rather than sloshing) in the center when the pan is nudged, 10 to 15 minutes (sometimes longer). Set the pan on a rack to cool. Remove the sides of the pan and transfer the tart to a platter. Serve the tart slightly warm or at room temperature. Refrigerate if not consuming within a few hours. The tart is best on the day it is made, but will keep in a covered container in the refrigerator for 2 or 3 days.

CHOCOLATE SABLÉS

These super-simple slice-and-bake shortbread cookies are only slightly sweet, but they have great cocoa flavor and a beautiful dark color. If you want fancier-looking cookies for a special occasion or to give as a gift (or something fun to do with children), you can roll the dough into balls, dredge with sugar, and flatten them individually before baking (see Note) or make the Chocolate-Mint Sandwich variation. Try the extra-bittersweet variation for even more chocolate flavor. **MAKES ABOUT 40 COOKIES**

---

1 cup plus 2 tablespoons (150 grams) teff flour

⅓ cup plus 1 tablespoon (60 grams) white rice flour

—OR—

½ cup plus 1 tablespoon (60 grams) Thai white rice flour

¼ cup plus 2 tablespoons (35 grams) natural unsweetened cocoa powder

⅔ cup (135 grams) sugar

Scant ½ teaspoon salt

⅛ teaspoon baking soda

12 tablespoons (1½ sticks/170 grams) unsalted butter, slightly softened and cut into chunks

¼ cup (60 grams) cream cheese

1 tablespoon water

1 teaspoon pure vanilla extract

EQUIPMENT

Food processor fitted with the steel blade (optional)

Baking sheets, lined with parchment paper

To make the dough by hand, put the teff and rice flours, cocoa, sugar, salt, and baking soda, in a large bowl and whisk until thoroughly blended. Add the butter chunks, cream cheese, water, and vanilla extract. Use a fork or the back of a large spoon to mash and mix the ingredients together until all are blended into a smooth, soft dough.

To make the dough in a food processor, put the teff and rice flours, cocoa, sugar, salt, and baking soda in the food processor. Pulse to mix. Add the butter chunks, cream cheese, water, and vanilla. Pulse until the mixture forms a smooth, soft dough. Scrape the bowl and blend in any stray flour at the bottom of the bowl with your fingers.

On a sheet of wax paper, shape the dough into a 10-inch log about 2 inches in diameter. Wrap the dough and refrigerate it for at least 2 hours, but preferably overnight.

Position racks in the upper and lower thirds of the oven. Preheat the oven to 325°F.

Slice the chilled cookie dough into ¼-inch slices and place them 1 inch apart on the lined sheets. Bake for 25 minutes, or until the cookies are firm to the touch (it is hard to tell by looking, but flip one cookie over and see if it is slightly browned on the bottom). Rotate the pans from front to back and top to bottom a little over halfway through the baking. Place the pans on racks or slide the liners from the pans onto racks to cool. Cool the cookies completely before stacking or storing. The cookies will keep for at least 2 weeks in an airtight container.

NOTE: If you have time to be a little fussier, you can shape chilled dough into 1-inch balls and roll them in sugar before placing them 2 inches apart on the lined baking sheets. Flatten each ball to about ¼ inch by covering with a piece wax paper and pressing with a flat-bottomed cup; peel off the wax paper and repeat with the remaining cookies. Bake as directed.

VARIATIONS

Spicy Chocolate Sablés

Add ¾ teaspoon ground cinnamon and ⅛ teaspoon each ground cayenne and freshly ground black pepper with the dry ingredients.

Extra-Bittersweet Chocolate Sablés

Put half of the sugar in the processor with 1½ ounces (45 grams) coarsely chopped unsweetened chocolate or a high-percentage (70% or higher) dark chocolate. Pulse until the chocolate pieces are the size of sesame seeds. Add the remaining ingredients and proceed as directed. These cookies take extra time to crisp up after they are cool because the chocolate takes longer to set and harden.

Chocolate-Mint Sandwich Cookies

Melt 4 ounces (115 grams) dark, milk, or white chocolate according to the instructions on page 35 and flavor with 2 drops (use an eye dropper!) of mint oil. Don't use mint extract or any non-oil-based flavoring or the chocolate will seize. Taste and adjust the flavor with additional drops of mint oil if necessary.

Sandwich cookies with ½ teaspoon of the minty chocolate. Let them stand to set the chocolate before serving.

COCOA CREPES FILLED WITH CHOCOLATE AND WALNUTS

Teff flour with a little cocoa powder makes delicate yet flavorful chocolate crepes to fill with chocolate, nuts, and cinnamon sugar. Add whipped cream or a little scoop of vanilla ice cream for a grand dessert, if you like. You could also fill crepes with Nutella (with or without bananas!) as described in the variation that follows, or make the Hungarian Crepe Cake (page 255). You can't go wrong. See page 21 for more about making crepes, including do-ahead tips and how to make crepes with other flours. **MAKES EIGHTEEN OR MORE 6-INCH CREPES TO SERVE 6**

FOR THE CREPES

1 cup (130 grams) teff flour

2 tablespoons (12 grams) unsweetened cocoa powder

¼ cup (50 grams) sugar

¼ teaspoon salt

2 tablespoons (30 grams) unsalted butter, melted, plus more for the pan

3 large eggs

1¼ cups whole milk, plus more as needed

¼ cup water

FOR THE FILLING

3 ounces (85 grams) 60% to 70% (or your choice) chocolate, finely chopped

½ cup (55 grams) finely chopped walnuts

Cinnamon sugar or plain sugar for sprinkling

Lightly sweetened whipped cream (see page 343), or 6 small scoops of vanilla ice cream (optional)

EQUIPMENT

6-inch crepe pan or skillet

In a medium bowl, whisk the flour, cocoa powder, sugar, salt, butter, and eggs with about one-quarter of the milk until blended and very smooth. Whisk in the remaining milk and the water. Cover and refrigerate the batter for at least an hour (to let the flour hydrate) and up to 2 days. Stir the batter well before and frequently as you use it.

Heat the pan over medium-high heat and brush lightly with butter. When a drop of water sizzles on the pan, lift the pan off the burner and pour in 2 tablespoons of the batter. Immediately tilt and rotate the pan, shaking as necessary to coat the bottom of the pan entirely. Fill in any holes with extra drops of batter. Set the pan back on the burner and cook until the surface of the crepe no longer looks wet and the underside is lightly browned in places, 30 to 60 seconds. Loosen the edges of the crepe with a spatula and flip it over with the spatula or (carefully) with your fingers. Cook 10 to 20 seconds longer. Slide or flip the crepe onto a piece of wax paper. Repeat with the remaining batter, buttering the pan only as necessary. If the crepes seem too thick, adjust the batter with a little extra milk. Crepes can be stacked as they come out of the pan—they won't stick to each other.

To assemble the crepes, sprinkle half of one crepe with a generous teaspoon each of chopped chocolate and walnuts, followed by a couple of pinches of cinnamon sugar. Fold the bare half of the crepe over the filling and fold again into quarters. Repeat with the remaining crepes. Crepes can be prepared to this point, covered, and refrigerated up to 1 day before serving.

(recipe continues)

To reheat and serve, arrange the filled crepes in a single layer on a parchment-lined baking sheet. Brush lightly with melted butter if you like and sprinkle with more cinnamon sugar. Reheat in a preheated 425°F oven on the upper rack for 5 to 10 minutes until hot. Overlap the crepes on a warmed serving platter, or plate 3 crepes per serving on warmed plates. Top each serving with ice cream if you like, or pass a bowl of whipped cream.

NOTE: If you have a helper, you can serve crepes informally in the kitchen, hot from the pan. The helper can assemble each crepe as you slide it out of the pan.

VARIATION: Nutella-Filled Teff Crepes (with or without Bananas)

Make the Cocoa Crepes as directed, but omit the chocolate and nut filling and spread each with a generous tablespoon of Nutella (with or without cinnamon sugar), and a few thin slices of banana, if serving immediately. Fold and serve at once, or cover, refrigerate, and reheat as directed.

HUNGARIAN CREPE CAKE

Here, rather than being folded, crepes are stacked with filling in between them, creating several ultrathin cake layers. You can wrap and stash the "cake" in the fridge up to a day ahead and simply reheat it before serving. This crepe cake was inspired by a recipe in *The Essence of Chocolate*, by Robert Steinberg and John Scharffenberger. **SERVES 8 TO 10**

---

5 ounces (140 grams) 60% to 70% chocolate (or your choice), broken or coarsely chopped

⅔ cup (95 grams) hazelnuts, toasted and skins rubbed off (see page 309)

¼ cup (50 grams) sugar

About 1 cup (12 ounces/340 grams) apricot preserves

20 to 25 8-inch Cocoa Crepes (page 253, double recipe, cooked in an 8-inch crepe pan)

EQUIPMENT

Food processor fitted with the steel blade

Pulse the chocolate, nuts, and sugar in the food processor until the nuts and chocolate are coarsely ground. Scrape the mixture into a small bowl. Put the preserves in the processor and pulse to puree any large pieces. Scrape into another bowl.

Place one crepe on a baking sheet (or on a sheet of wax paper, if making ahead). Spread a very thin layer of apricot preserves, about 2 teaspoons, over the crepe. Sprinkle with a tablespoon of the chocolate mixture. Spread a second crepe with jam and place it on top of the first, then sprinkle it with the chocolate mixture as before. Repeat with all of the remaining crepes, leaving the top one bare.

The cake can be made to this point, wrapped airtight, and refrigerated for up to 1 day, or frozen for up to 3 months and defrosted before baking. To serve, cover the top of the cake with a greased piece of foil and bake in a preheated 350°F oven for 20 to 30 minutes, just until heated through. Transfer to a serving platter and serve hot or warm.

BITTERSWEET TEFF BROWNIES

These moist and deeply chocolate brownies have a light, rather elegant melt-in-your-mouth texture. Teff flour has a nuance of cocoa flavor to start with, so it is a natural choice for brownies. If you need something dressier than brownies, bake the batter in a 9-inch round pan and serve wedges with whipped cream—and perhaps a scattering of seasonal berries—and call it dessert. Either way, the recipe comes together quickly and the results remain deliciously moist for a few days. **MAKES SIXTEEN 2-INCH BROWNIES**

---

10 tablespoons (1¼ sticks/140 grams) unsalted butter, cut into chunks

6 ounces (170 grams) 70% chocolate, coarsely chopped

1 scant cup (185 grams) sugar

¾ cup (100 grams) teff flour

¼ teaspoon salt

1 teaspoon pure vanilla extract (optional)

3 large eggs, cold

1 cup (100 grams) walnut or pecan pieces (optional)

EQUIPMENT

Handheld mixer

8-inch square pan, bottom and all four sides lined with foil

Position a rack in the lower third of the oven and preheat the oven to 350°F.

Melt the butter with the chocolate in a medium heatproof bowl set directly in a wide skillet of barely simmering water. Stir frequently until the mixture is melted and smooth.

Remove the bowl from the water and cool the mixture to lukewarm. Stir in the sugar, teff flour, salt, and vanilla, if using. Add all of the eggs and beat on high speed with the handheld mixer for about 2 minutes. The batter will get thicker and a little lighter in color, like chocolate frosting. Stir in the nuts, if using.

Scrape the batter into the pan and spread it evenly. Bake for 30 to 35 minutes, until a toothpick inserted in the center comes out fairly dry and clean (don't worry; the brownies will be moist even if the toothpick is not).

Cool on a rack. Lift the foil ends to transfer the brownies to a cutting board. Cut into 16 squares. The brownies may be kept in an airtight container for 2 to 3 days.

VARIATION: Cocoa Teff Brownies

Cocoa brownies have a softer texture than chocolate brownies. Substitute ¾ cup (65 grams) unsweetened cocoa powder for the chocolate. Increase the butter to 13 tablespoons (185 grams), and increase the sugar to 1 cup plus 3 tablespoons (235 grams).

TEFF BLINI

Like a cross between pancakes and English muffins, these are crispy at the edges and spongy and chewy in the middle. As with the Better Than Buckwheat Blini on page 185, you will love them for breakfast or tea with butter and honey, but don't forget to try them with caviar or smoked salmon with chopped red onions and sour cream, as described on page 186. **SERVES 8 AS AN APPETIZER**

---

3 tablespoons warm water (105°F to 115°F)

1 teaspoon active dry yeast

½ teaspoon sugar

1 cup plus 1 tablespoon (160 grams) white rice flour

—OR—

1½ cups (160 grams) Thai white rice flour

½ cup plus 2 tablespoons (80 grams) teff flour

½ cup (80 grams) glutinous rice flour

—OR—

⅔ cup (80 grams) Thai glutinous rice flour

2 tablespoons plain yogurt (any percent fat) or slightly watered down Greek yogurt

1 teaspoon salt

1½ cups water

Soybean or corn oil, for frying

½ cup (170 grams) honey

4 tablespoons (½ stick/55 grams) unsalted butter, melted

EQUIPMENT

12-inch frying pan

Combine the warm water, yeast, and sugar in a large bowl and let rest for 5 minutes. Add the flours, yogurt, salt, and 1½ cups water and whisk until smooth. Cover the bowl with plastic wrap and set over a slightly smaller bowl filled with warm water—the water can touch or not. Let rest for about half an hour. The batter should be fairly thick; if it is not, whisk in a little extra rice flour.

Heat a large frying pan over medium heat, pour in 1 tablespoon of the oil, and lift and tilt the pan to spread the oil. Whisk or stir the batter to mix thoroughly, and spoon tablespoonfuls of batter onto the pan, leaving about an inch between pancakes.

Cook until lightly browned on the first side and bubbles break on the top surface. Use a spatula to turn the pancakes and cook the second side until lightly browned. Repeat with the remaining batter, adding oil to the pan before each batch. Serve immediately or keep warm in a covered dish. To serve, drizzle with honey and butter. Leftovers may be refrigerated for up to 2 days and reheated for 5 minutes in a 400°F oven.

TANGY AROMATIC CRACKERS

It's hard for me to choose a favorite between these and the Seed Crackers on page 163—they are both stellar in their own way and worth the time it takes to produce them. These have enough personality to partner with aged cheeses or rich pâté. **MAKES 1½ DOZEN LARGE CRACKERS / PHOTO ON PAGE 195**

½ cup (80 grams) brown rice flour

½ cup (80 grams) white rice flour

—OR—

¾ cup (80 grams) Thai white rice flour

¼ cup plus 2 tablespoons (40 grams) gluten-free oat flour

½ cup plus 2 tablespoons (80 grams) teff flour

¼ cup plus 2 tablespoons (40 grams) flaxseed meal, or ¼ cup (40 grams) whole flaxseeds, finely ground

⅔ cup (90 grams) unsalted pumpkin seeds, raw or roasted

1 teaspoon fennel seeds

1 teaspoon caraway seeds

1 teaspoon cumin seeds

1 tablespoon packed brown sugar

1 teaspoon salt

¾ cup plus 1 tablespoon water

1 tablespoon balsamic vinegar

2 teaspoons baking powder

¼ cup extra-virgin olive oil

Sea salt or other coarse salt for sprinkling (about 1 tablespoon)

EQUIPMENT

Stand mixer with paddle attachment

Baking sheets

Rolling pin

Position racks in the upper and lower thirds of the oven and preheat the oven to 450°F.

Mix the flours, flaxseed meal, pumpkin, fennel, caraway, and cumin seeds, brown sugar, and salt in the bowl of the stand mixer fitted with the paddle attachment. Add the water and vinegar and beat for 2 minutes on medium speed to form a very thick, sticky dough that might wrap around the paddle at first. Sprinkle in the baking powder, add the olive oil, and beat for 1 minute on medium speed to thoroughly incorporate the oil.

Cut four pieces of parchment the size of a baking sheet. Drop three 2-tablespoon lumps of dough evenly spaced down the length of a parchment sheet. Cover with another piece of parchment and roll the dough into oblongs about 3 by 5 inches (or larger) and a scant ⅛ inch thick. Peel off the top parchment (save for reuse), sprinkle the dough very lightly with coarse salt, and place the parchment with the crackers *dough side down* on a baking sheet.

Bake for 5 to 6 minutes, or until the crackers are browned at the edges. Remove the pan from the oven and carefully peel off the parchment (save for reuse). Flip the crackers over with a spatula and return them to the oven for 2 to 3 minutes, or until well browned at the edges. Repeat with the remaining dough, baking two pans at a time and rotating them from upper to lower and front to back for even baking. While the crackers are baking, continue to roll out more dough; as soon as a pan of crackers is done, flip the next batch onto the baking sheet (it's okay if the sheets are still hot as long as you put them in the oven immediately).

Cool the crackers thoroughly on parchment or a rack before storing in an airtight container for up to 10 days. The crackers may be refreshed before serving by baking for 5 minutes at 400°F.

CHAPTER SEVEN

SORGHUM FLOUR

THE FIRST sorghum seeds probably traveled across the Atlantic with African slaves. Although most Americans have likely never tried it, sorghum is now the fifth most important cereal crop in the world, and the United States is the biggest producer. Drought-resistant sorghum is still a subsistence crop in Africa and South America, but we cultivate sorghum mostly for livestock, fuel, or export. Novels of the old South—romantic and otherwise—mention the molasses-like sorghum syrup used as a sweetener in cooking or like honey on biscuits or corn bread. It's incredibly delicious! But this flavorful syrup has long since been replaced by cheaper corn syrup and cane sugar: sorghum syrup is now a specialty product, hardly known outside of the American South. As for the grain itself, we Americans don't eat much sorghum—yet.

But sorghum may be poised for a revival, even greatness. It tastes good, and it's a nutritious whole grain—actually an "ancient" grain—that has already gained traction as a niche product in the gluten-free community. Meanwhile, sorghum cultivation is expanding in this country: sorghum needs one-third less water than corn and grows in soils inhospitable to most other crops.

Sorghum flour is a pale sandy tan color with a mild, decidedly sweet aroma, slightly reminiscent of cornmeal but without the bitter note of the latter. Sorghum is often described as having a neutral flavor and thus is a good substitute for wheat flour. I find that it has a gentle but distinctive flavor with nuances of wheat, corn, or even oats, depending on the recipe. Neutral reputation aside, it makes a more definite statement than does rice flour, so it is less useful as a background player, and more useful as a partner for complementary ingredients (see sidebar). You will notice that the flavor of sorghum flour goes remarkably well with other Southern flavors and ingredients, such as bourbon, pecans, and even ham!

Alone, sorghum flour makes beautiful pancakes, crepes, and waffles (see page 21) with no need to add other flours or even xanthan gum to hold it together. A simple sorghum sponge cake (or génoise) will become an important new basic in the baking repertoire. Sorghum makes delicious shortbread, banana muffins, and scones with figs and aniseed. You'll also find an easy rustic ice cream made with sorghum flour and topped with salted peanuts caramelized in sorghum syrup (see page 292).

FLAVOR AFFINITIES FOR SORGHUM FLOUR

Butter, berries, pecans, peanuts, dates, figs, banana, and warm spices such as ginger, cinnamon, cardamom, aniseed, nutmeg

WHERE TO BUY AND HOW TO STORE

Sorghum flour is available in better supermarkets in the specialty flour section of the baking aisle, or by mail order from Authentic Foods (extra-fine sorghum flour) or Bob's Red Mill (see Resources, page 351). Sorghum flour is a whole grain and should be stored in an airtight container, away from heat and light, for 2 to 3 months at room temperature, or 6 months in the refrigerator, and up to 12 months in the freezer.

SORGHUM LAYER CAKE WITH WALNUT PRALINE BUTTERCREAM

This is a subtle but dressy little layer cake, more European in style than American. The thin layers of sponge cake are made with sorghum flour and moistened with bourbon (or rum or brandy) syrup and filled with a new version of classic French buttercream laced with crushed caramelized walnuts. You can make the cake, caramelized walnuts, and buttercream over several days to lighten the load and, once assembled, the cake should rest for a day to allow the flavors and textures to mellow. Review the Tips for Working with Buttercream (page 349), and keep in mind that it is much richer than ordinary frosting and therefore should be spread very thinly, as directed. For a quicker, more casual, but still delicious cake, try the variation filled and frosted with Praline Whipped Cream (page 345) instead of buttercream. **SERVES 10 TO 12**

FOR THE CAKE

5 tablespoons (70 grams) Clarified Butter (page 335) or ghee (see page 38)

¾ cup (90 grams) sorghum flour

3 tablespoons (35 grams) white rice flour
NOTE: This recipe is not successful with Thai white rice flour

⅔ cup (130 grams) sugar

4 large eggs

⅛ teaspoon salt

FOR THE CAKE ASSEMBLY AND FINISHING

2 tablespoons (25 grams) sugar

1 tablespoon plus 1 teaspoon water

2 tablespoons bourbon (or brandy or rum), or to taste

2 cups New Classic Buttercream (page 348), made with egg yolks

⅔ cup (90 grams) Walnut Praline Brittle powder (see page 337)

Position a rack in the lower third of the oven and preheat the oven to 350°F. Line the bottom of the pan with parchment paper, but do not grease the sides of the pan.

To make the cake, put the clarified butter in a small pot or microwavable container ready to reheat when needed, and have a 4- to 5-cup bowl ready to pour it into as well—the bowl must be big enough to allow you to fold some batter into the butter later.

Whisk the sorghum and rice flours and 2 tablespoons of the sugar together thoroughly in a medium bowl.

Combine the remaining sugar, eggs, and salt in the bowl of the stand mixer and beat with the whisk attachment on high speed for 4 to 5 minutes. The mixture should be light colored and tripled in volume, and you should see well-defined tracks as the whisk spins; when the whisk is lifted, the mixture should fall in a thick, fluffy rope that dissolves slowly on the surface of the batter.

Just before the eggs are ready, heat the clarified butter until very hot and pour it into the reserved bowl.

Remove the bowl from the mixer. Sift one-third of the flour over the eggs. Fold with a large rubber spatula until the flour is almost blended into the batter. Repeat with half of the remaining flour. Fold in the rest of the flour. Scrape about one-quarter of the

EQUIPMENT

8-by-2-inch round cake pan

Stand mixer with whisk attachment

Sifter or medium-mesh strainer

Pastry brush

batter into the hot butter. Fold until blended. Scrape the buttery batter over the remaining batter and fold just until blended. Scrape the batter into the pan.

Bake for 30 to 35 minutes, until the cake is golden brown on top. It will have puffed up and then settled level, but it won't have pulled away from the sides of the pan and a toothpick inserted in the center should come out clean and dry. Set the pan on a rack to cool.

At your convenience (the cake can be warm or completely cool), run a small spatula around the inside of the pan, pressing against the sides of the pan to avoid tearing the cake. Invert the pan onto a rack to remove the cake and peel off the parchment liner. Turn the cake right side up to finish cooling before filling, frosting, or storing. The cake may be wrapped airtight and stored at room temperature for 2 days, or frozen for up to 3 months.

To assemble the cake, combine the sugar, water, and bourbon in a small jar. Stir and let the mixture stand until the sugar is dissolved. Cover until needed.

Mix the buttercream with the Walnut Praline powder.

If the cake is not level, use a serrated knife to trim the edges. Cut the cake horizontally into 3 thin layers. If the layers are too delicate to pick up without breaking, use a rimless baking sheet or a flexible plastic cutting mat to lift and transfer them. Place the first layer on an 8-inch cardboard round or the base of an 8-inch springform pan to support the bottom. Brush the layer with up to a third of the bourbon syrup—don't soak it, just make it nicely damp. Spread ½ cup of the buttercream over the layer. Moisten the second cake layer with less bourbon syrup than before (since it will also be moistened on the other side) and place it moist side down on the first layer. Moisten the top of the layer similarly and spread with another ½ cup of buttercream. Moisten the third cake layer with bourbon syrup and put it moist side down on the second layer. Brush the top with the remaining bourbon syrup. Level the cake by pressing gently with the bottom of a pan, making sure the assembly is not at a tilt—press it gently from the sides as necessary to straighten it. Cover and refrigerate the remaining buttercream. Wrap the cake in plastic and refrigerate it for at least 2 hours and up to 24 hours.

To finish the cake, remove the reserved buttercream from the fridge and soften it according to the instructions on page 349. Unwrap the cake and trim the sides evenly with a serrated knife. Put less than half of the remaining buttercream in a separate bowl (to avoid getting crumbs in it) and spread it as thinly as possible over the top and sides of the cake just to smooth and glue on any loose crumbs—this is the "crumb coat" and it won't be pretty. Chill the cake for 15 minutes to set the buttercream. Frost the top and sides of the cake with just enough of the remaining buttercream to make a thin, smooth layer. Put the cake in a covered container or under a dome and return it to the refrigerator. Remove it from the fridge at least an hour before serving to soften the buttercream and bring out the flavors and aromas. Leftover cake keeps in the refrigerator for a few days.

NOTE: Sorghum flour, like some other whole grains, produces a sponge cake with a slightly grainy texture. Such a cake might seem even coarser if paired with a supersmooth filling or frosting. The great trick here is teaming the cake with an element that has even more texture. The fine crunch from the caramelized nuts balances the texture of the cake perfectly—and the flavors all work together beautifully, so the results are both delicious and unexpectedly refined.

VARIATIONS

Sorghum Layer Cake with Coffee-Walnut Praline Buttercream

Dissolve 1 to 1½ teaspoons instant espresso powder in ¼ teaspoon water and mix it into the buttercream with the praline powder.

Sorghum Layer Cake with Praline Whipped Cream

Make the cake as directed but cut it into 2 instead of 3 layers. Moisten the layers with bourbon syrup as directed. Omit the buttercream. Fill and frost the cake generously with 1½ batches (3 cups or more) of Praline Whipped Cream (page 345) made with any nut praline you fancy. Refrigerate it for several hours before serving. Remove from the refrigerator about 30 minutes before serving.

BANANA MUFFINS

Sorghum has a gentle, not-too-assertive flavor that balances perfectly with sweet ripe bananas. These sturdy, satisfying little muffins are great for breakfast or a snack on the go. **MAKES 12 MUFFINS**

⅔ cup (135 grams) packed brown sugar

1 large egg

⅓ cup vegetable oil

1 cup (150 grams) white rice flour
—OR—
1½ cups (150 grams) Thai white rice flour

¼ cup plus 2 tablespoons (50 grams) sorghum flour

½ teaspoon baking soda

1½ teaspoons baking powder

¼ teaspoon salt

1 cup mashed ripe bananas

⅔ cup (70 grams) coarsely chopped walnuts

EQUIPMENT

Stand mixer with paddle attachment or handheld mixer

12-cup muffin pan, lined with paper liners

Position a rack in the lower third of the oven and preheat the oven to 350°F.

Combine the sugar, egg, and oil in the bowl of the stand mixer and beat with the paddle attachment on medium speed until lighter in color, about 2 minutes. Or beat in a large bowl with the handheld mixer on medium-high speed for 3 to 4 minutes.

Add the rice and sorghum flours, baking soda, baking powder, salt, and bananas and beat on low speed until smooth. Beat in the walnuts. Scoop the mixture into the prepared muffin pan.

Bake for 18 to 20 minutes, until a toothpick inserted in the center comes out clean. Set the pan on a rack to cool.

The muffins keep in an airtight container at room temperature for 3 days.

FIG AND ANISE SCONES

Sorghum flour has an understated, almost-but-not-quite corn flavor that makes a great match for the assertive dried figs and aromatic spice here. Scones have the best texture and rise if you allow the dough to rest in the refrigerator for at least 2 hours before baking, either as a whole log or sliced. **MAKES TWELVE 3-INCH SCONES**

1⅓ cups (200 grams) white rice flour (preferably superfine)

—OR—

2 cups (200 grams) Thai white rice flour

Scant ½ cup (60 grams) sorghum flour

¼ cup (50 grams) granulated sugar

1 teaspoon aniseed

¼ teaspoon xanthan gum

1 tablespoon baking powder

½ teaspoon salt

1 cup heavy cream

½ cup plain yogurt (any percent fat) or slightly watered down Greek yogurt

⅔ cup (120 grams) stemmed, coarsely chopped dried figs

About 2 tablespoons coarse sugar, such as turbinado, for sprinkling

EQUIPMENT

Stand mixer with paddle attachment

Baking sheet, lined with parchment paper

Combine the rice and sorghum flours, granulated sugar, aniseed, xanthan gum, baking powder, salt, cream, and yogurt in the bowl of the stand mixer and beat with the paddle attachment for 2 minutes; the dough will be very stiff. It is important to beat the dough enough or the scones won't rise well; don't worry about overbeating. Beat in the figs.

Form the dough into a log 2 inches in diameter, wrap it in plastic, and refrigerate for at least 2 hours or up to 2 days.

Position a rack in the upper third of the oven and preheat the oven to 400°F.

Cut the log into 12 thick slices. Place the slices 2 inches apart on the lined pan and sprinkle with the coarse sugar. Bake for 20 to 25 minutes, or until the scones are browned on top and bottom. Serve immediately or cool on a rack and toast before serving.

STRAWBERRY TARTLETS

Tartlets are a little fussier to make than full-size tarts, but they are delightful to serve. You can have your way with them: major in fruit by using only a little pastry cream and covering each tartlet completely with whole berries—or go for broke with lots of pastry cream and just a garnish of fruit, as shown in the photo. SERVES 7

A double recipe of dough from Sorghum Pecan Tart (page 277), made with an additional 2 tablespoons of sugar

2 to 3 teaspoons unsalted butter, very soft, to moisture-proof the crusts (if necessary)

1 cup The New Vanilla Pastry Cream (page 342)

1 pint (8 ounces/225 grams) ripe strawberries (or other berries or a combination of berries)

EQUIPMENT

Seven 4½-inch fluted tartlet pans with removable bottoms

Rimmed baking sheet

Grease the tartlet pans with vegetable oil spray or butter.

Make the tart dough as directed and divide it into 7 equal portions (about 70 grams each). The dough will be very soft. Use your fingers to press it evenly into the bottoms of the tartlet pans; don't worry about the sides. Lay a piece of plastic wrap over one pan and press a straight-sided, flat-bottomed cup (such as a measuring cup) firmly into the center, pressing the dough flat and pushing it up against the sides of the pan. Work the cup around the bottom and the sides and into the corners of the pan until the bottom and sides are evenly lined; use your fingers to finish the sides and square off the top edges as necessary, and leave the plastic wrap in place. Repeat with the remaining pans. Refrigerate them for at least 2 hours, but preferably overnight and up to 3 days.

Position a rack in the lower third of the oven and preheat the oven to 325° F.

Remove the plastic from the tartlet shells and place them on the baking sheet. Bake for 15 minutes. Check to see if the dough has puffed up; use the back of a spoon to press it gently back down and against the sides of the pan if necessary. Bake for 15 to 20 minutes longer, until the crusts are a rich golden brown all over (if they are too pale, they won't be as delicious as they should be) and the dough has pulled away from the sides of the pan. Set the baking sheet on a rack and let the crusts cool completely before removing them or filling them. The shells may be prepared to this point, wrapped airtight, and kept at room temperature for up to 1 week.

To assemble the tartlets, push up on the bottoms of the pans to loosen the sides of the crust, but leave the crusts in the pans for support. If you will not be serving the tartlets within a couple of hours, moisture-proof the crusts by smearing the thinnest possible coat of softened butter over the bottom of each one: your finger is the best tool for doing this. Chill the crusts briefly to set the butter.

Fill each crust with 1 or 2 tablespoons of pastry cream. Arrange whole or cut strawberries, or other whole berries, over the pastry cream. If not serving within a couple of hours, refrigerate the tartlets in a covered container. See the box on page 276 to add a little shine to the tart.

To serve, push the bottom of the pans up and remove the sides. Carefully loosen the crust from the pan bottom by slipping a very thin knife blade between them. Use a spatula to lift or slide the tartlets onto serving plates.

STRAWBERRY-BLUEBERRY TART

A buttery tender shortbread crust with fresh ripe fruit on a thin bed of pastry cream is one of the best summer desserts of all. If you do not have any memories of eating such tarts in France, you may want to double the amount of pastry cream under the fruit, but for me the minimal amount of pastry cream is just as it should be! The tart is best on the day it is assembled, but you can make the pastry cream and the crust a day or two ahead: keep the former covered in the fridge and the latter wrapped airtight at room temperature. This tart is also excellent with an oat flour crust: substitute ¾ cup (75 grams) oat flour for the sorghum flour. SERVES 8

FOR THE CRUST

½ cup plus 1 tablespoon (75 grams) sorghum flour

3 tablespoons (30 grams) white rice flour
—OR—
scant ⅓ cup (30 grams) Thai white rice flour

¼ cup (50 grams) sugar

¼ teaspoon salt

1/16 teaspoon baking soda

6 tablespoons (85 grams) unsalted butter, slightly softened and cut into chunks

2 tablespoons (30 grams) cream cheese

1½ teaspoons water

½ teaspoon pure vanilla extract

About 2 teaspoons unsalted butter, very soft, to moisture-proof the crust (if necessary)

FOR THE FILLING

1 cup (or 2 cups) The New Vanilla Pastry Cream (page 342)

2 pints (16 ounces/450 grams) ripe strawberries, rinsed (see Note), hulled, and halved if large

½ pint (5 ounces/140 grams) blueberries

To make the crust, follow the directions in Sorghum Pecan Tart (page 277).

Refrigerate the pan for at least 2 hours, but preferably overnight and up to 3 days.

Place the pan on the baking sheet and peel off the plastic wrap. Bake for 30 to 35 minutes, checking after 15 to 20 minutes. If the crust has puffed up on the bottom, press it back down carefully with the back of a fork. Continue baking until the crust is golden brown and has slightly pulled away from the edges of the pan all around. Set the pan on a rack to cool completely before filling.

To assemble the tart, leave the tart shell in the pan for support. If you will not be serving the tart within a couple of hours, moisture-proof the crust by spreading the bottom with the thinnest possible layer of soft butter; use a flexible plastic spreader or even your fingers to smear the butter over the crust. Chill the crust to set the butter before adding the pastry cream.

Spread the pastry cream evenly in the crust. Start arranging the strawberries around the edges of the tart and work toward the middle. Arrange whole berries as close together as possible, or arrange halved berries cut side up and overlapping. Toss the blueberries between the strawberries to cover any exposed pastry cream. Refrigerate unless serving within 2 hours. To serve, remove the sides of the pan and transfer the tart to a platter. The tart is best on the day it is assembled.

(recipe continues)

EQUIPMENT

Food processor fitted with the steel blade (optional)

9½-inch fluted tart pan with a removable bottom

Rimmed baking sheet

NOTE: Wet fruit is one of the worst things that can happen to a fresh fruit tart! After rinsing the berries, spread them in one layer on a dish towel or on layers of paper towels. Make sure they are completely dry before assembling the tart—if necessary, pat them with more towels, set them in front of a fan, or dry them with a hairdryer on the cool setting.

Extra Shine

If you want a little shine to the fruit, add it shortly before serving: Simmer ¼ cup red currant jelly with 2 teaspoons sugar for a minute or two until the mixture is thick and sticky. Use a pastry brush to brush or dab the fruit with just enough of the glaze to make the fruit sparkle—don't expect to use it all. Leftover glaze can be covered and refrigerated for another tart, spread on toast, or even swirled into yogurt.

SORGHUM PECAN TART

This sweet seasonal favorite gets a light but truly superb flavor makeover as a tart with a sorghum flour crust and, if you like, a little sorghum syrup in the filling. You should make this tart at least a day ahead. Having "rediscovered" a piece of it seven days after I made it, I was amazed at how the pecan flavor had developed and all of the flavors were more integrated; a very good tart became an excellent tart and even more convenient to make during the busy holidays! **SERVES 10**

FOR THE CRUST

½ cup plus 1 tablespoon (75 grams) sorghum flour

3 tablespoons (30 grams) white rice flour
—OR—
scant ⅓ cup (30 grams) Thai white rice flour

3 tablespoons (35 grams) sugar

¼ teaspoon salt

1/16 teaspoon baking soda

6 tablespoons (85 grams) unsalted butter, slightly softened and cut into chunks

2 tablespoons (30 grams) cream cheese

1½ teaspoons water

½ teaspoon pure vanilla extract

FOR THE FILLING

¼ cup light corn syrup (90 grams), or 2 tablespoons light corn syrup and 2 tablespoons sorghum syrup

1 tablespoon unsalted butter

¾ cup (170 grams) packed dark brown sugar

¼ teaspoon salt

1 tablespoon rum (or bourbon or brandy)

1 teaspoon pure vanilla extract

3 large eggs

2 cups (7 ounces/200 grams) pecan halves

Grease the tart pan with vegetable oil spray or butter.

To make the crust by hand, put the sorghum and rice flours, sugar, salt, and baking soda in a medium bowl and whisk until thoroughly blended. Add the butter chunks, cream cheese, water, and vanilla. Use a fork or the back of a large spoon to mash and mix the ingredients together until all are blended into a smooth, soft dough.

To make the crust in a food processor, put the sorghum and rice flours, sugar, salt, and baking soda in the food processor. Pulse to blend. Add the butter chunks, cream cheese, water, and vanilla. Pulse until the mixture forms a smooth, soft dough. Scrape the bowl and blend in any stray flour at the bottom of the bowl with your fingers.

Transfer the dough to the tart pan.

The dough may seem much softer than other tart doughs. Use the heel of your hand and then your fingers and/or a small offset spatula to spread the dough all over the bottom of the pan. Press it squarely into the corners with the side of your index finger to prevent extra thickness at the bottom edge, and press it as evenly as possible up the sides of the pan, squaring it off along the top edge. Have patience; there is just enough dough (although you may not think so at first). If there is too much dough in one place (or hiding in the corners), pinch or scrape it off and move it elsewhere. Spread or smear it smooth with the spatula. Here's a final trick for a perfectly even crust: Press a sheet of plastic wrap against the bottom and sides of the pan and lay a paper towel on top. Set a straight-sided, flat-bottomed cup on the towel; press and slide the cup all over the bottom and into the corners to smooth and even

Lightly sweetened whipped cream (see page 343)

EQUIPMENT

9½-inch fluted tart pan with removable bottom

Food processor fitted with the steel blade (optional)

Rimmed baking sheet

the surface. Leave the plastic wrap in place. Refrigerate the pan for at least 2 hours, but preferably overnight and up to 3 days.

Position a rack in the lower third of the oven and preheat the oven to 325°F. Spread the pecans on a baking sheet and bake for 6 to 9 minutes, moving them around the sheet once or twice, until fragrant and lightly colored. Set aside.

Peel off the plastic wrap and place the tart pan on the rimmed baking sheet. Bake the crust for 30 to 35 minutes, checking after 15 to 20 minutes. If the crust has puffed up on the bottom or sides, press it back in place carefully with the back of a fork. Continue baking until the crust looks fully done—golden to deep golden brown.

Make the filling while the crust is baking: Combine the corn syrup (or corn and sorghum syrups) and butter in a medium stainless steel bowl set directly in a skillet of barely simmering water (or in the top of a double boiler over barely simmering water) and stir until the butter is melted. Stir in the brown sugar, salt, rum, and vanilla. Add the eggs and whisk until the mixture is well blended and hot to the touch. Set the whole water bath (or double boiler) aside, stirring the filling from time to time, until needed. When the crust is baked, remove it from the oven, but leave the oven on. Scatter the pecans over the crust. Whisk the hot filling to reblend it. Pour the filling over the nuts and turn any upside-down nuts right side up if you like.

Return the tart to the oven. Bake for 15 to 20 minutes, or until the filling is slightly puffed and feels firm but springy when pressed with your finger. Cool the tart on a rack. Remove the sides of the pan and transfer the tart to a platter. Serve at room temperature with lightly sweetened whipped cream. The tart keeps, covered, at room temperature for up to a week.

TOASTY PECAN BISCOTTI WITH HIBISCUS PEARS

This is a gorgeous and unusual composed dessert: spicy, nut-studded cookies served with a sweet, intensely flavorful garnet-hued compote of pears poached in dried hibiscus-flower syrup. The cookies are technically biscotti since they are baked twice (once on each side), and have all the crunch of regular biscotti, but their nutty coating makes them extra crunchy and even more irresistible. Dried hibiscus flowers (flor de Jamaica) are available in Latin grocery stores and tea shops. You will also love the biscotti with Silky Chocolate Pudding (page 80). **SERVES 8**

⅔ cup (100 grams) white rice flour

—OR—

1 cup (100 grams) Thai white rice flour

¼ cup plus 2 tablespoons (50 grams) sorghum flour

¼ cup (50 grams) granulated sugar

⅛ teaspoon xanthan gum

1 teaspoon baking powder

¼ teaspoon salt

¼ cup plus 2 tablespoons milk (any percent fat)

8 tablespoons (1 stick/115 grams) unsalted butter, softened

¼ cup (50 grams) coarse sugar, such as turbinado

½ teaspoon cinnamon (preferably canela cinnamon)

2 cups (200 grams) coarsely chopped pecans

Hibiscus Pears (recipe follows)

EQUIPMENT

Stand mixer with paddle attachment

Baking sheet, lined with parchment paper

Rolling pin

Combine the rice and sorghum flours, granulated sugar, xanthan gum, baking powder, salt, and milk in the bowl of the stand mixer and beat with the paddle attachment for 2 minutes; the dough will be very stiff. It is important to beat the dough long enough or the biscotti won't hold together; don't worry about overbeating. Add the butter and beat until thoroughly incorporated. Scrape the dough into a flat patty, wrap in plastic, and refrigerate for about 2 hours, until firm.

Position a rack in the upper third of the oven and preheat the oven to 400°F.

Mix the coarse sugar, cinnamon, and pecans in a bowl and scoop about half of it onto the parchment paper on the baking sheet. Turn the dough patty out onto the mixture and turn and press it to cover both sides with nuts. Place it in the center of the paper and cover with another piece of parchment. Roll out to an 8-by-10-inch rectangle about ½ inch thick. Peel off the top sheet of parchment and pat or roll the remaining nut mixture onto the dough. Using a long thin knife, cut the dough straight down to make 2 sections, each 5 by 8 inches, and then cut each section into 8 bars, 1 inch wide and 5 inches long. Wipe the knife with a paper towel as needed. Use the knife or a spatula to move the bars so that they are evenly spaced on the same sheet. Press any stray nuts into the bars. Bake for 15 minutes. Remove from the oven and turn the biscotti over with a spatula, moving the browner ones to the center of the pan. Return to the oven and bake for 8 to 10 minutes, until the biscotti are lightly browned on top and bottom. Set the pan on a rack to cool.

To serve, spoon ⅓ to ½ cup of the pears into individual bowls with some of their syrup, accompanied by a couple of biscotti.

(recipe continues)

Hibiscus Pears

MAKES 1 QUART

¼ cup (15 grams) crushed dried hibiscus flowers (flor de Jamaica) or loose hibiscus tea, or ½ cup (15 grams) whole dried hibiscus flowers

1¼ cups water

1¼ cups (250 grams) sugar

2 tablespoons lemon juice

2 pounds (910 grams) firm pears (to yield about 4 cups fruit)

EQUIPMENT

Strainer

Basket-type coffee filter or paper towel

NOTE: Any leftover syrup makes a great cocktail, such as 2 parts syrup to 1 part gin over ice.

Place the hibiscus in a medium heatproof bowl. Bring the water to a boil in a medium nonreactive saucepan and pour the water over the hibiscus. Let steep for 10 minutes. Line the strainer with the coffee filter or paper towel and pour the liquid back into the saucepan; squeeze the hibiscus in the filter to extract all of the juice and discard the hibiscus. Add the sugar and lemon juice and bring to a simmer over medium heat.

Peel, quarter, and core the pears, then cut each quarter into 4 long slices. Add the pears to the simmering syrup. Simmer for 8 to 10 minutes, or until translucent all the way through, turning carefully once or twice. Cover the pan and let cool for about an hour. Refrigerate the pears in the syrup for at least 2 hours. They may be stored, covered, in the refrigerator for up to 1 week.

NUTMEG SHORTBREAD

When I grated a little nutmeg over a stray piece of broken sorghum tart crust—on a whim, really—I found a heavenly combination. If nutmeg is one of your favorite flavors, you'll love these. They definitely belong on a holiday cookie tray, and do serve them with the eggnog. Do-ahead for best flavor and texture: Let the dough rest and hydrate overnight in the fridge before baking, and store cookies for at least 1 day before serving. **MAKES 40 COOKIES**

---

1 cup plus 2 tablespoons (150 grams) sorghum flour

⅓ cup plus 1 tablespoon (60 grams) white rice flour
—OR—
½ cup plus 1 tablespoon (60 grams) Thai white rice flour

½ cup (100 grams) sugar

¾ teaspoon freshly grated nutmeg

Rounded ¼ teaspoon salt

12 tablespoons (1½ sticks/170 grams) unsalted butter, slightly softened and cut into chunks

¼ cup (60 grams) cream cheese

1 tablespoon water

1 teaspoon pure vanilla extract

Nutmeg sugar: 1 tablespoon sugar mixed with ½ teaspoon grated nutmeg

EQUIPMENT

Food processor fitted with the steel blade (optional)

Baking sheets, lined with parchment paper

To make the dough by hand, put the sorghum and rice flours, ½ cup sugar, ¾ teaspoon nutmeg, and salt in a large bowl and whisk until thoroughly blended. Add the butter chunks, cream cheese, water, and vanilla. Use a fork or the back of a large spoon to mash and mix the ingredients together until all are blended into a smooth, soft dough.

To make the dough in a food processor, put the sorghum and rice flours, ½ cup sugar, ¾ teaspoon nutmeg, and salt in the food processor and pulse to mix. Add the butter chunks, cream cheese, water, and vanilla. Pulse until the mixture forms a smooth, soft dough. Scrape the bowl and mix in any stray flour at the bottom of the bowl with your fingers.

On a sheet of wax paper set on a baking sheet, press the dough into an even 8-by-10-inch rectangle about ½ inch thick. Cover the dough with plastic wrap and refrigerate it for at least 2 hours and up to 3 days.

Position racks in the upper and lower thirds of the oven. Preheat the oven to 325°F.

Sprinkle the dough with the nutmeg sugar. Cut the dough lengthwise into 1-inch strips, then crosswise into 2-inch lengths and place them an inch apart on the lined pans. Bake for 20 to 25 minutes, until the cookies are golden brown at the edges. Rotate the pans from front to back and top to bottom a little over halfway through the baking. Place the pans on racks, or slide the liners from the pans onto racks to cool. Cool the cookies completely before stacking or storing. The cookies will keep for at least 2 weeks in an airtight container.

SALTED PEANUT SHORTIES

These tender melt-in-your-mouth shortbread cookies are loaded with salted peanuts and scented with bourbon. Do-ahead for best flavor and texture: Let the dough rest and hydrate overnight in the fridge before baking, and store cookies for at least 1 day before serving. For a special treat, sandwich them just before serving with purchased dulce de leche. **MAKES 45 TO 50 COOKIES**

---

2 cups (225 grams) roasted or dry-roasted salted peanuts

1 cup plus 2 tablespoons (150 grams) sorghum flour

⅓ cup plus 1 tablespoon (60 grams) white rice flour
—OR—
½ cup plus 1 tablespoon (60 grams) Thai white rice flour

⅔ cup (135 grams) sugar

12 tablespoons (1½ sticks/170 grams) unsalted butter, slightly softened and cut into chunks

¼ cup (60 grams) cream cheese

2 tablespoons bourbon or 1 tablespoon water

1 teaspoon pure vanilla extract

EQUIPMENT

Food processor fitted with the steel blade

Baking sheets, lined with parchment paper

Put the peanuts, sorghum and rice flours, and sugar in the food processor. Pulse until the peanuts are mostly pulverized, but not too finely. Add the butter chunks, cream cheese, bourbon or water, and vanilla. Pulse until the mixture forms a smooth, soft dough. Scrape the bowl and blend in any stray flour at the bottom of the bowl with your fingers. On a sheet of wax paper, shape the dough into a 12-inch log about 2 inches in diameter. Wrap the dough and refrigerate it for at least 2 hours, but preferably overnight.

Position racks in the upper and lower thirds of the oven and preheat the oven to 325°F.

Slice the chilled cookie dough into ¼-inch slices and place them 1 inch apart on the lined sheets. Bake for 20 to 25 minutes, until the cookies are golden brown at the edges. Rotate the pans from front to back and top to bottom a little over halfway through the baking. Place the pans on racks, or slide the liners from the pans onto racks to cool. Cool the cookies completely before stacking or storing. The cookies will keep for at least 2 weeks in an airtight container.

VARIATIONS

Coffee Walnut Shorties

Substitute a generous cup (115 grams) walnuts, 1 teaspoon finely ground coffee beans (regular, not espresso roast), and ½ teaspoon salt for the peanuts. Use water instead of bourbon.

Nibby Walnut Shorties

Substitute a generous cup (115 grams) walnuts and ½ teaspoon salt for the peanuts. Process only until the nuts are chopped medium-fine instead of pulverized. Add ¼ cup roasted cacao nibs with the remaining ingredients. Use water instead of bourbon.

GINGER-PEACH SQUARES

These festive cookies are both tenderly crunchy and a little chewy from the dried fruit and candied ginger filling. You can sprinkle some chopped pecans onto the dough with the filling to make fruit- and nut-filled squares instead: measurements are not critical so you can do this by eye. **MAKES TWENTY-FIVE 2-INCH COOKIES**

---

1 cup plus 2 tablespoons (150 grams) sorghum flour

⅓ cup plus 1 tablespoon (60 grams) white rice flour
—OR—
½ cup plus 1 tablespoon (60 grams) Thai white rice flour

½ cup (100 grams) granulated sugar

Rounded ¼ teaspoon salt

12 tablespoons (1½ sticks/170 grams) unsalted butter, slightly softened and cut into chunks

¼ cup (60 grams) cream cheese

1 tablespoon water

1 teaspoon pure vanilla extract

1 cup (140 grams) diced dried peaches

3 tablespoons (30 grams) finely diced candied ginger

2 to 3 tablespoons coarse sugar, such as turbinado, for sprinkling

Whole nutmeg (optional, for grating)

EQUIPMENT

Food processor fitted with the steel blade (optional)

Baking sheets, lined with parchment paper

To make the dough by hand, put the sorghum and rice flours, granulated sugar, and salt in a large bowl and whisk until thoroughly blended. Add the butter chunks, cream cheese, water, and vanilla. Use a fork or the back of a large spoon to mash and mix the ingredients together until all are blended into a smooth, soft dough.

To make the dough in a food processor, put the sorghum and rice flours, granulated sugar, and salt in the food processor and pulse to mix. Add the butter chunks, cream cheese, water, and vanilla. Pulse until the mixture forms a smooth, soft dough. Scrape the bowl and blend in any stray flour at the bottom of the bowl with your fingers.

Divide the dough in half. Use a dark pencil or a marker to draw a 10-by-10-inch square on each of two pieces of parchment paper. Turn one sheet upside down (to prevent the dough from touching the pencil or ink marks) on the counter and anchor the corners with tape.

Press and then spread (with a small offset spatula) one piece of the dough to make an even ¼-inch layer within the square. Check to be sure the center is not thicker than the edges. Remove the tape and slide the parchment onto a sheet pan. Distribute the peach pieces evenly over the dough, followed by the ginger pieces. Set aside.

Turn the second parchment sheet over and secure it to the counter. Spread the remaining dough over it as before and release the tape. Cover the fruit-topped dough with the second sheet of dough as follows: Place the pan with the fruit-topped dough next to the plain dough. Lift the far edge of the parchment under the plain dough until the dough dangles over the counter. Line up the dangling edge with the far edge of the fruit-topped dough. Let the edges touch, and then lower the dough sheet toward you, to cover the fruit. (It's easier to do than to describe!) Don't peel the paper

from the dough; just press very gently all over to adhere the dough and fruit. Put the baking sheet in the fridge and refrigerate for at least 2 hours, but preferably overnight.

Position racks in the upper and lower thirds of the oven and preheat the oven to 325°F.

Remove the dough from the refrigerator and peel the paper from the top. Dust the top of the dough and a cutting board very lightly with rice flour. Lift the parchment under the dough, flip the dough over onto the cutting board, and peel off the paper. Sprinkle with the coarse sugar and pat lightly to make sure the sugar adheres. Use a heavy knife to trim the edges. Use a straight-down "guillotine" stroke to cut 5 strips and then cut each strip into 5 pieces to make 25 pieces. Don't worry if the dough cracks when you cut it. Use a spatula to lift and place the cookies 1 inch apart on the lined pans.

Bake for 15 to 20 minutes, until the cookies are golden brown at the edges and deep golden brown when you peek underneath (carefully, as the cookies are very fragile while hot). Rotate the pans from front to back and top to bottom a little over halfway through the baking. Place the pans on racks, or slide the liners from the pans onto racks to cool. Cool the cookies completely before stacking or storing. The cookies may be kept for at least a week in an airtight container.

NUTELLA SANDWICH COOKIES

Shortbread cookies filled with chocolate-hazelnut spread—what's not to love? You can add ground hazelnuts to the cookie dough or not, as you like. **MAKES ABOUT 25 SANDWICH COOKIES**

---

1 cup plus 2 tablespoons (150 grams) sorghum flour

⅓ cup plus 1 tablespoon (60 grams) white rice flour

—OR—

½ cup plus 1 tablespoon (60 grams) Thai white rice flour

⅔ cup (130 grams) granulated sugar

½ teaspoon salt

1 cup (140 grams) hazelnuts (optional)

12 tablespoons (1½ sticks/170 grams) unsalted butter, slightly softened, cut into chunks

¼ cup (60 grams) cream cheese

1 tablespoon water

1 teaspoon pure vanilla extract

Coarse sugar, such as turbinado, for rolling

¾ cup chocolate-hazelnut spread, such as Nutella

EQUIPMENT

Food processor fitted with the steel blade

Baking sheets, lined with parchment paper

Combine the sorghum and rice flours, granulated sugar, salt, and hazelnuts, if using, in the food processor and pulse until the hazelnuts are finely ground. Add the butter chunks, cream cheese, water, and vanilla. Pulse until the mixture forms a smooth, soft dough. Scrape the bowl and blend in any stray flour at the bottom of the bowl with your fingers. On a sheet of wax paper, shape the dough into a 12-inch log about 2 inches in diameter. Wrap the dough and refrigerate it for at least 2 hours, but preferably overnight.

Position racks in the upper and lower thirds of the oven and preheat the oven to 325°F.

Roll the chilled cookie dough in the coarse sugar, pressing it to adhere. Cut into slices less than ¼ inch thick and place them 1 inch apart on the lined sheets. Bake for 20 to 25 minutes, until the cookies are golden brown at the edges. Rotate the pans from front to back and top to bottom a little over halfway through the baking. Place the pans on racks, or slide the liners from the pans onto racks to cool. Cool the cookies completely before stacking, filling, or storing. Unfilled cookies may be kept for at least 2 weeks in an airtight container.

Turn half of the cookies upside down. Spoon the chocolate-hazelnut spread into one corner of a resealable plastic freezer bag. Clip about ¼ inch from the corner and pipe about 1½ teaspoons onto each upside-down cookie. Cover with a right-side-up cookie and press very gently to spread the filling toward the edges. The filled cookies will keep in an airtight container for up to 3 days, although they will soften after the first few hours.

SORGHUM CINNAMON STICKS

Imagine superlight stick-shaped biscotti dredged in spicy sugar. These are addictive and noisy to eat. The subtle honey/maple flavor actually comes from the sorghum flour itself. If you like Indian spices or just want to add a touch of complexity to the flavors here, add a pinch of garam masala or even pumpkin pie spice to the cinnamon sugar. That's all! **MAKES 24 STICKS**

---

3 tablespoons (45 grams) unsalted butter

¾ cup plus 1 tablespoon (110 grams) sorghum flour

¼ cup plus 2 tablespoons (75 grams) sugar

Generous ⅛ teaspoon salt

½ teaspoon baking powder

2 large eggs, at room temperature

Cinnamon sugar: 2 tablespoons sugar mixed with ½ teaspoon cinnamon (optional: a pinch of garam masala)

EQUIPMENT

Skillet at least 12 inches wide

Baking sheets, 1 lined with parchment paper or foil, 1 unlined

Stand mixer with whisk attachment, or handheld mixer

8-inch square pan, bottom lined with parchment paper

Melt the butter in a large heavy-bottomed skillet. Take the pan off the heat, add the sorghum flour, and stir to coat all of the flour grains with butter. The mixture will have the consistency of slightly damp sand. Return the skillet to the stove and cook over medium-high heat, stirring constantly with a heatproof spatula or fork; scrape the bottom and sides of the pan, turning the flour, and spread or rake to redistribute it continuously so that it toasts evenly. Continue to cook and stir until the mixture colors slightly and smells toasted; it may begin to smoke a little. Toasting the flour will take 4 to 6 minutes. Scrape the flour onto the lined baking sheet and spread it out to cool while preheating the oven.

Position a rack in the lower third of the oven and preheat the oven to 350°F.

Combine the sugar, salt, baking powder, and eggs in the bowl of the stand mixer fitted with the whisk attachment (or in a large mixing bowl if using a handheld mixer). Beat on high speed for 3 to 5 minutes, until thick and light. Poke and mash any large lumps in the toasted flour and then pour it over the egg mixture. Fold just until evenly mixed. Scrape the batter into the lined baking pan and spread it evenly; it will be a thin layer only about ½ inch deep.

Bake for 20 to 25 minutes, or until golden brown and springy to the touch. Set the pan on a rack to cool. Leave the oven on, but reduce the temperature to 300°F.

Slide a slim knife around the edges of the pan to detach the baked sheet. Invert the pan onto a rack and peel off the liner. Place the baked sheet right side up on a cutting board. Cut it in half with a sharp serrated knife. Cut each half crosswise into slices a scant ¾ inch wide.

Put the cinnamon sugar in a shallow dish and gently dredge the sticks liberally on all sides. Arrange the slices slightly apart, standing up, on an unlined baking sheet. Bake for 20 to 25 minutes, or until slightly golden brown. Rotate the sheet from front to back about halfway through the baking time. Cool the cookies completely before storing. The cookies may be stored in an airtight container for several weeks.

NOTE: If you want to double the recipe, use a 9-by-13-inch baking pan. Toasting the sorghum flour may take 10 to 12 minutes and require a skillet at least 14 inches wide (wider is better). The larger amount of flour may clump while toasting; if necessary, mash out lumps with a fork while the flour is cooling. It may be easier to toast the flour in two batches instead, and then proceed as directed.

SORGHUM ICE CREAM WITH PEANUT BRITTLE

Sorghum syrup tastes like a kinder and gentler version of molasses, with hints of toffee and coffee. Once a staple sweetener in the rural South, it's now a specialty item worth seeking out and using as you might use honey or molasses: serve it with biscuits or toasted English muffins and butter, or drizzle it on ice cream or yogurt or even goat cheese! Once you've got it in the cupboard, why not use it to sweeten and flavor this simple ice cream *and* its peanut brittle topping? Sorghum flour replaces the usual cornstarch in this traditional egg-free Sicilian-style gelato. **MAKES ABOUT 1 QUART**

FOR THE ICE CREAM

3 tablespoons (25 grams) sorghum flour

¼ cup (50 grams) sugar

Generous ¼ teaspoon salt

2⅔ cups half-and-half

1 cup whole milk

⅓ cup sorghum syrup

FOR THE BRITTLE

½ cup (100 grams) sugar

¼ cup sorghum syrup

1 cup (115 grams) roasted or dry-roasted salted peanuts

EQUIPMENT

Baking sheet, lined with parchment paper

Ice cream maker

For the ice cream, put the sorghum flour, sugar, and salt in a medium saucepan and whisk in enough of the half-and-half to make a smooth paste. Whisk in the remaining half-and-half and the milk. Cook over medium heat, sweeping the bottom, sides, and corners of the pot to prevent scorching, until the mixture comes to a simmer. Continue to cook and stir, adjusting the heat to maintain a lively simmer, for 3 minutes to cook the flour completely. Whisk in the sorghum syrup. Scrape into a bowl and cool at room temperature or in the refrigerator.

Meanwhile, make the peanut brittle: Place the lined baking sheet near the stove. Combine the sugar, sorghum syrup, and peanuts in a medium saucepan. Cook over medium heat, stirring with a heatproof silicone spatula, until the nuts are coated with syrup. Continue stirring the nuts, scraping the sides and corners of the pan, as the bubbling syrup begins to resemble slightly foamy lava. The syrup will darken and begin to smoke; continue stirring until the syrup is a deep mahogany color, but not burnt. Scrape the nuts onto the parchment and spread them out. As soon as the brittle is cool enough to handle, but still quite warm, break it into pieces and transfer it (still warm) to a zipper-lock bag or airtight container to prevent it from getting sticky. It will keep for at least a month.

Freeze the ice cream base according to the instructions with your ice cream maker. Store the ice cream in the freezer until needed. If the ice cream is rock hard, let it soften in the fridge for 15 minutes or more before serving or in the microwave on Defrost for a few seconds at a time until scoopable. Serve with chopped and broken brittle. The ice cream is best within 2 or 3 days.

CHAPTER EIGHT

NUT AND COCONUT FLOURS

THE SO-CALLED nut flours—ground or grated nuts—and various forms of dried coconut are grouped together in this chapter because they are all composed of fat and fiber, they are not starchy like grain flours, and they behave similarly in recipes. Moreover, they are often combined in recipes because they taste good together.

Nut flour, also called nut meal, is nothing more than finely grated nuts that may or may not have been blanched first to remove their skins. Nut flour is finer and fluffier, and absorbs more moisture from any batter, than the nut meal that you can make with the steel blade in your food processor or blender. Often, they can be used interchangeably (with different but excellent results). A few recipes actually start by pulverizing the nuts in the processor with the steel blade before adding the remaining ingredients. You could substitute an equal weight of purchased nut flour instead of starting with whole nuts, without harm, but why bother if the recipe is made in the processor anyway? For recipes that specifically call for nut flour, use purchased flour, or make it yourself (see page 355).

Nuts are, well, nutty. Their slight bitterness is balanced by rich and sweet aspects. Nut flours and meals, as well as dried shredded coconut, lend a pebbly texture to finished baked goods as well as contribute their characteristic delicious flavors. Paired with rice flour in a recipe, their texture can also counteract rice flour's tendency to produce gritty cookies or gummy cakes. Paired with coarse flours, nut flours can make cakes feel softer and finer on the palate and cookies more delicate. Nut flours and dried shredded coconut work beautifully in myriad recipes, as a main flour, a partner, or an accent.

Coconut flour is finely ground dried unsweetened coconut with about 80 percent of its fat removed, leaving it very high in fiber, with much less fat than regular dried coconut. It has a gentle coconut flavor and aroma.

Coconut flour is tricky (and very non-intuitive) to work with. Because it absorbs enormous quantities of available moisture from any batter, it can turn cakes or cookies into either cardboard or mush! It is often used more as a functional rather than a flavor ingredient—to add fiber and retain moisture—in gluten-free baking. In this book, it is used along with dried shredded coconut or nuts to boost coconut flavor and crunch without adding unnecessary extra fat. Crunchy Coconut Cookies (page 321) and several tarts with coconut crusts are sensational examples.

FLAVOR AFFINITIES FOR NUT AND COCONUT FLOURS

Coconut Flour:

Tropical flavors, brown sugar, coffee, nuts, dark chocolate

Almond Flour/Meal:

Apricots, plums, cherries, peaches, dark chocolate

Walnut and Pecan Flours:

Coffee, brown sugar, caramel, any chocolate, blackberries, bourbon, rum, dates

Hazelnut Flour:

Dark fruit, pears, milk and dark chocolate

WHERE TO BUY AND HOW TO STORE

Like whole-grain flours, nut flours are prone to rancidity. Whether purchased or homemade, they should be stored in an airtight container, away from heat and light, for 2 to 3 months at room temperature, or 6 months in the refrigerator, and at least 12 months in the freezer. See Resources (page 351) for information on where to buy. Dried shredded coconut, flaked coconut, and coconut flour keep well in an airtight container at room temperature for at least 12 months.

NUTTY SPONGE CAKE

This cake is moist and deliciously nutty, yet relatively light in texture. Irresistible! You can make it with any type of nut flour (other than coconut or chestnut), and you can switch out the grated zest, or add a few drops of orange or rose flower water, or ground spices, and create a dozen new sponge cakes with one recipe. This cake is always good plain, or with berries and whipped cream. **SERVES 10 TO 12**

---

5 large eggs, separated, at room temperature

1 teaspoon pure vanilla extract

¼ teaspoon almond extract

Grated zest of ½ medium lemon

¼ teaspoon salt

⅔ cup (130 grams) sugar

¼ teaspoon cream of tartar

2¾ cups (10 ounces/280 grams) almond flour/meal (see Note)

Lightly sweetened whipped cream (see page 343; optional)

Berries, plain or sweetened (optional)

EQUIPMENT

10-inch tube pan with removable bottom

Stand mixer with whisk attachment, or handheld mixer

NOTE: Nut flours vary in weight per cup. To substitute another type of nut flour, measure by weight (280 grams) rather than cups. Omit the almond extract if not using almond flour.

Position a rack in the lower third of the oven. Preheat the oven to 325 °F. Grease the pan with vegetable oil spray or butter.

In a large mixing bowl, whisk the egg yolks, vanilla, almond extract, lemon zest, salt, and ⅓ cup (65 grams) of the sugar until thick and pale yellow.

Combine the egg whites and cream of tartar in the bowl of the stand mixer (or another large bowl if using a handheld mixer) and beat with the whisk attachment on medium-high speed (or on high speed with the handheld mixer) until the egg whites are creamy white and hold a soft shape when the beaters are lifted. Gradually beat in the remaining ⅓ cup sugar, at high speed, until peaks are stiff but not dry when the whisk is lifted.

Scrape half of the egg whites over the yolks and pour half of the nut flour on top. Use a large rubber spatula to fold until the elements are partially incorporated. Repeat with the remaining egg whites and nut flour, folding just until incorporated. Scrape the batter into the pan (it will be not quite half full) and spread it evenly. Bake until the cake is golden brown on top, springs back when lightly pressed, and a toothpick or bamboo skewer inserted in the center comes out clean, 30 to 35 minutes. Set the pan on a rack to cool.

To unmold the cake, run a skewer around the tube and slide a thin spatula around the sides of the pan to detach the cake. Lift the tube to remove the cake. Slide the spatula under the cake all around to detach the bottom. Use two spatulas to lift the cake onto a serving platter or a rack. Serve the cake upside down or right side up. The cake keeps, wrapped airtight, at room temperature for at least 3 days, or in the freezer for up to 3 months; bring to room temperature before serving. Slice with a serrated knife.

ALMOND BUTTER CAKE WITH SPICED APRICOTS

Almond flour gives this fine-textured buttery cake a subtle nuance of almond flavor and fragrance—just enough in my book, but feel free to add ¼ teaspoon of almond extract for a bolder statement. Serve it sweet and pretty topped with a scoop of ice cream or make it modern with a crumble of goat cheese on top. Either way, there's just enough pepper in the apricots to catch your attention without making them too spicy to handle. In lieu of apricots, you can serve the cake with sliced fresh oranges and Cardamom Brittle (page 338) or with nothing at all! **SERVES 12 TO 16**

2⅓ cups (360 grams) white rice flour
—OR—
3½ cups (360 grams) Thai white rice flour

1 cup (100 grams) almond flour/meal

2 cups minus 3 tablespoons (360 grams) sugar

½ pound (2 sticks/225 grams) unsalted butter, very soft

¾ teaspoon salt

2 teaspoons baking powder

1 teaspoon baking soda

½ teaspoon xanthan gum

1 cup plain yogurt (any percent fat) or slightly watered down Greek yogurt

4 large eggs

2 teaspoons pure vanilla extract

Spiced Apricots (recipe follows)

7 ounces (200 grams) fresh goat cheese, or 1 quart vanilla ice cream (optional)

EQUIPMENT

Two 9-by-2-inch round cake pans or two 8½-by-4½-inch (6-cup) loaf pans

Stand mixer with paddle attachment

Position a rack in the lower third of the oven and preheat the oven to 350°F. Grease the sides of the round pans with vegetable oil spray or butter and line the bottoms with parchment. Line the bottom and all four sides of the loaf pans with parchment.

Combine the rice and almond flours, sugar, butter, and salt in the bowl of the stand mixer and beat on medium speed with the paddle attachment until the mixture is the texture of brown sugar, about a minute. Add the baking powder, baking soda, xanthan gum, yogurt, eggs, and vanilla and beat on medium-high speed for 2 to 3 minutes; the batter should be very smooth and fluffy. Scrape the batter into the prepared pans and bake the round cakes for 25 to 30 minutes, the loaves for 50 to 55 minutes, until a toothpick or bamboo skewer inserted in the center comes out clean (the loaves will be deep golden brown on top). Cool the cakes in the pans on a rack.

Slide a slim knife or small metal spatula around the edges of each round cake to detach it from the pan. Invert the cakes onto the rack and peel off the paper liner. Turn the cakes right side up. Simply tip the loaves out of their pans and set them right side up.

To serve round cakes, slice wedges and place in shallow bowls with 4 or 5 apricots and some of their syrup. Crumble about a tablespoon of goat cheese on top of each serving, or top with a small scoop of ice cream, if desired. To serve loaves, arrange slices on a serving platter. Chop the apricots roughly and pile into a bowl with a little of their juice; let guests help themselves.

The plain cake keeps, wrapped airtight, for up to 3 days at room temperature.

(recipe continues)

Spiced Apricots

MAKES ALMOST 4 CUPS

1 pound (455 grams) plump whole dried apricots (see Note)

½ cup commercial balsamic vinegar (nothing fancy or expensive is necessary)

½ cup (170 grams) honey

8 whole cloves

1 teaspoon black peppercorns

2-inch cinnamon stick, crumbled

½ cup water

NOTE: Dried whole apricots are plumper and sweeter than halves, and they work best in this recipe. If you can only get dried apricot halves, simmer them gently in a little water to soften them before proceeding. Taste and add a little sugar if necessary toward the end.

Combine the apricots, vinegar, honey, cloves, peppercorns, cinnamon, and water in a small nonreactive saucepan. Bring to a simmer, cover, and remove from the heat. Let steep for 30 minutes. Chill before serving. Serve the apricots whole in their syrup (strained to remove the spices and cinnamon stick) as a compote, or roughly chopped, as a condiment. The apricots will keep in a covered container in the refrigerator for up to 1 week.

COCONUT CHIFFON CAKE

Tender, moist, and light, this cake has lovely flavors of rice and coconut. The batter is gorgeous, fluffy, and light to start with, and using a large bowl will make folding easier. Serve cake wedges with slices of fresh pineapple, plain, grilled, or even sautéed in butter with a little sugar for browning, and pass a bowl of chocolate sauce. Or splash the cake with a little rum and top with whipped cream and grated lime zest. **SERVES 10**

½ cup plus 3 tablespoons (140 grams) sugar

2 large egg yolks, at room temperature

⅜ cup coconut water or light or low-fat Asian coconut milk

¼ cup flavorless vegetable oil, such as corn or safflower

1 teaspoon pure vanilla extract

⅓ cup plus 1 tablespoon (35 grams) unsweetened dried shredded coconut

⅔ cup (100 grams) white rice flour
NOTE: This recipe is not successful with Thai white rice flour.

1½ teaspoons baking powder

¼ teaspoon salt

4 large egg whites, at room temperature

¼ teaspoon cream of tartar

EQUIPMENT

9-by-2-inch round cake pan

Stand mixer with whisk attachment

Position a rack in the lower third of the oven and preheat the oven to 325°F. Line the bottom of the pan with parchment; leave the sides ungreased.

Set aside 2 tablespoons (25 grams) of the sugar for later.

Add the remaining sugar to a large mixing bowl with the egg yolks, coconut water or milk, oil, vanilla, coconut, rice flour, baking powder, and salt. Whisk briskly until the mixture is well blended. Set aside.

In the bowl of the stand mixer, beat the egg whites and cream of tartar with the whisk attachment on medium-high speed until the egg whites are creamy white and hold a soft shape when the beaters are lifted. Slowly sprinkle in the reserved 2 tablespoons sugar, beating on high speed until the egg whites are stiff but not dry. Scrape one-quarter of the egg whites onto the batter and fold them in with a rubber spatula. Fold in the remaining egg whites. Scrape the batter into the prepared pan and spread it evenly.

Bake for 28 to 33 minutes, until the top of the cake is golden brown and springs back when you press it gently with a finger and a toothpick inserted in the center of the cake comes out clean and dry.

Set the pan on a rack to cool for about 10 minutes. Slide a thin knife or a small metal spatula around the sides of the cake to detach it from the pan. Invert the cake onto the rack and peel off the parchment liner. Turn the cake right side up on the rack to cool completely. Once cool, the cake keeps, wrapped airtight, at room temperature for 3 days, or in the freezer for up to 3 months; bring to room temperature before serving.

HAZELNUT LAYER CAKE WITH DARK CHOCOLATE FROSTING AND BLACKBERRY PRESERVES

Earthy sweet and bitter chocolate with buttery hazelnuts is a legendary partnership. Blackberry preserves are a perfect bright counterpoint to them in this simple layer cake. **SERVES 12 TO 16**

2⅓ cups (360 grams) white rice flour
—OR—
3½ cups (360 grams) Thai white rice flour

1¼ cups (100 grams) hazelnut flour

2 cups minus 3 tablespoons (360 grams) sugar

½ pound (2 sticks/225 grams) unsalted butter, very soft

¾ teaspoon salt

2 teaspoons baking powder

1 teaspoon baking soda

½ teaspoon xanthan gum

1 cup plain yogurt (any percent fat) or slightly watered down Greek yogurt

4 large eggs

2 teaspoons pure vanilla extract

½ cup seedless blackberry preserves or fruit spread

½ recipe Dark Chocolate Frosting (page 95)

EQUIPMENT

Two 9-by-2-inch round cake pans

Stand mixer with paddle attachment

Position a rack in the lower third of the oven and preheat the oven to 350°F. Grease the sides of the pans with vegetable oil spray or butter, and line the bottoms with parchment.

Combine the rice and hazelnut flours, sugar, butter, and salt in the bowl of the stand mixer and beat on medium speed with the paddle attachment until the mixture is the texture of brown sugar, about a minute. Add the baking powder, baking soda, xanthan gum, yogurt, eggs, and vanilla and beat on medium-high speed for 2 to 3 minutes; the batter should be very smooth and fluffy. Scrape the batter into the prepared pans and bake for 25 to 30 minutes, until a toothpick inserted in the center comes out clean. Cool the cakes in the pans on a rack.

Slide a slim knife or small metal spatula around the edges of each cake to detach it from the pan. Invert the cakes onto a rack and peel off the paper liner. Turn the cakes right side up.

When the cakes are cool, invert one layer onto a serving plate. Spread with the blackberry preserves and set the second layer, right side up, on top. Spread about a third of the frosting in a very thin layer all over the cake to smooth the surfaces and glue on any crumbs. Refrigerate for about 30 minutes to set the frosting. Stir the remaining frosting until smooth (warm it slightly if necessary) and frost the cake as luxuriously as you want. Depending on how much frosting you like, there may be some left over. The cake keeps in an airtight container for up to 5 days in the refrigerator. Bring to room temperature before serving.

PISTACHIO BUTTER CAKE WITH SWEET PICKLED PEACHES

Pistachios give this cake a toasty, mysterious flavor and a touch of color. Spicy peach or fig compote adds to the exotic vibe, and any leftovers are good with a mellow aged Cheddar cheese or spooned over some vanilla ice cream. **SERVES 12 TO 16**

2⅓ cups (360 grams) white rice flour

—OR—

3½ cups (360 grams) Thai white rice flour

1 cup (100 grams) finely ground pistachios or pistachio flour

2 cups minus 3 tablespoons (360 grams) sugar

½ pound (2 sticks/225 grams) unsalted butter, very soft

¾ teaspoon salt

2 teaspoons baking powder

1 teaspoon baking soda

½ teaspoon xanthan gum

1 cup plain yogurt (any percent fat) or slightly watered down Greek yogurt

4 large eggs

2 teaspoons pure vanilla extract

Sweet Pickled Peaches (recipe follows)

Tangy Whipped Cream (page 345)

EQUIPMENT

Two 9-by-2-inch round cake pans

Stand mixer with paddle attachment

Position a rack in the lower third of the oven and preheat the oven to 350°F. Grease the sides of the pans with vegetable oil spray or butter, and line the bottoms with parchment.

Combine the rice flour, pistachios, sugar, butter, and salt in the bowl of the stand mixer and beat on medium speed with the paddle attachment until the mixture is the texture of brown sugar, about a minute. Add the baking powder, baking soda, xanthan gum, yogurt, eggs, and vanilla and beat on medium-high speed for 2 to 3 minutes; the batter should be very smooth and fluffy. Scrape the batter into the prepared pans and bake for 25 to 30 minutes, until a toothpick inserted in the center comes out clean. Cool the cakes in the pans on a rack.

Slide a slim knife or small metal spatula around the edges of each cake to detach it from the pan. Invert the cakes onto a rack and peel off the paper liners. Turn the cakes right side up.

Serve slices in shallow bowls with 6 to 8 peach halves and a tablespoon of syrup. Top with a dollop of whipped cream.

The cake keeps, wrapped airtight, for up to 3 days at room temperature.

(recipe continues)

Sweet Pickled Peaches

MAKES ABOUT 4 CUPS

1 pound (455 grams) plump dried peaches, halved or quartered (see Note)

½ cup white balsamic or other white wine vinegar

¾ cup (150 grams) sugar

2-inch cinnamon stick, crumbled

2 slices of fresh ginger

½ teaspoon mustard seeds

½ teaspoon allspice berries

½ teaspoon black peppercorns

½ teaspoon whole cloves

½ teaspoon coriander seeds

½ teaspoon crushed red pepper flakes

1 cup water

NOTE: If the peaches are dry and hard, simmer them in a little water to soften them before proceeding with the recipe.

Combine all of the ingredients in a nonreactive saucepan. Bring to a simmer, cover, and remove from the heat. Let steep for 30 minutes. Chill for at least 2 hours before serving. Serve with a little of the syrup (strained to eliminate the spices and cinnamon stick). The peaches will keep in a covered container in the refrigerator for up to 1 week.

VARIATION: Sweet Pickled Figs

Use 1 pound (455 grams) dried figs, stemmed and halved, instead of the peaches, increase the vinegar to 1 cup, and decrease the sugar to ½ cup (100 grams).

CARROT COCONUT ALMOND TORTE

Fragrant with lemon zest and cinnamon or cardamom, this moist, nubbly-textured European-style torte is an appealing—and quite different—alternative to the more familiar and beloved American Carrot Spice Cake (page 101) with its cream cheese frosting. Here, ground almonds and dried coconut take the place of flour, and the cake is leavened entirely with whipped egg whites instead of baking powder. Whipped cream or crème fraîche rather than frosting is the best topping for this one. Make the torte a day ahead for the best flavor. SERVES 10 TO 12

1½ cups (140 grams) almond flour/meal

½ cup (45 grams) unsweetened dried shredded coconut

2 cups (225 grams) lightly packed finely grated carrots (see Note)

1 small to medium lemon

4 large eggs, separated, at room temperature

¼ teaspoon salt

½ teaspoon (slightly rounded) ground cinnamon or cardamom

¼ teaspoon pure almond extract

1 cup (200 grams) sugar

¼ teaspoon cream of tartar

Unsweetened or lightly sweetened whipped cream (see page 343) or Whipped Crème Fraîche (page 345)

EQUIPMENT

8-by-3-inch springform pan or cheesecake pan with removable bottom

Stand mixer with whisk attachment, or handheld mixer

Position a rack in the lower third of the oven and preheat the oven to 325°F. Grease or spray the pan with vegetable oil spray.

In a medium bowl, mix the almond flour with the coconut. Set aside.

Stack four paper towels on the counter. A handful at a time, squeeze the grated carrots hard, over the sink or a bowl (if you plan to sip the juice), to extract as much juice as you can, then place them in the center of the paper towels. When all of the carrots have been squeezed once, gather the edges of the paper towels around the carrots and squeeze again—letting the towels absorb more juice. Set the bundle on the counter, place your hands on it, and lean with your full body weight to extract any remaining juice. Set aside.

Use a Microplane zester to grate the zest of the lemon directly into a large mixing bowl. Add the egg yolks, salt, cinnamon, and almond extract. Set aside ¼ cup (50 grams) of the sugar. Add the remaining sugar to the bowl. Whisk briskly for a minute or so until the mixture lightens in color. Sprinkle the carrots into the bowl, but don't mix them in.

Combine the egg whites and cream of tartar in the bowl of the stand mixer (or in another large bowl if using a handheld mixer). Beat with the whisk attachment on medium-high speed (or on high speed with the handheld mixer) until the egg whites are creamy white and hold a soft shape when the beaters are lifted. Slowly sprinkle in the reserved ¼ cup sugar, beating on high speed until the egg whites are stiff but not dry.

(recipe continues)

Scrape one-third of the egg whites on top of the batter. Use a large rubber spatula to fold the egg whites and carrots almost completely into the batter. Scrape the remaining egg whites into the bowl and pour the almond-coconut mixture over and around them. Fold just until the ingredients look evenly blended. Scrape the batter into the prepared pan and spread the surface level.

Bake for 50 to 55 minutes, until the surface of the torte is golden brown, the torte is beginning to separate from the sides of the pan, and it springs back when you press it gently with your fingers.

Cool the torte in the pan on a rack. Slide a slim knife or spatula around the sides to detach it from the pan, then remove the sides of the pan. Slide a spatula or knife between the cake and the bottom of the pan to detach it. Use a metal pancake turner to transfer the cake to a serving plate. Slice and serve with dollops of whipped cream. Leftover cake keeps for 3 or 4 days stored in a covered container at room temperature.

NOTE: Flavorful and well-squeezed carrots are key to producing a torte that is moist but not wet or soggy. The carrots must be finely grated or you will not be able to squeeze much juice from them. Use the fine shredding disk on your food processor, the smallest of the three shredding discs on a Mouli grater, or the finest grating surface (other than the one designed to zest citrus peels) on your box grater.

How to Toast Nuts

Even when chopped toasted nuts or nut flour is called for, toast whole nuts (or pecan or walnut halves) first and then chop or grind them. Toast pine nuts in a skillet as for coconut (see below); spread other nuts in a single loose layer on an ungreased baking sheet. Bake in a preheated oven at 350°F (for almonds and hazelnuts) or 325°F (for pecans and walnuts) for 10 to 20 minutes, depending on the type of nut and whether they are whole, halves, sliced, or slivered to start with. Check the color and flavor of the nuts from time to time and stir to redistribute them on the pan. Almonds and hazelnuts are done when they are golden brown inside. Pecans and walnuts are done when fragrant and lightly colored. To remove the skins from toasted hazelnuts, let them cool thoroughly, then rub them together in your hands or in a tea towel or place them in a large coarse-mesh strainer and rub them against the mesh to coax off the skins. Some skins will always remain; don't worry.

How to Toast Coconut (or Pine Nuts or Sesame Seeds)

Have a sheet of parchment or a bowl near the stove before you start. Spread the coconut in a dry skillet over medium heat. Stir constantly until the coconut begins to color slightly. Turn the heat down (once hot, coconut burns quickly) and continue to stir until evenly light golden brown, then evenly brown—take the pan off the heat a bit early and continue to toss and stir, letting the residual heat of the pan finish toasting the coconut slowly and evenly. Scrape the coconut onto the parchment or into the bowl and let cool before using.

COCONUT CHOCOLATE PECAN TORTE

Light, moist, and slightly gooey like a macaroon laced with bits of dark chocolate, this cake is a terrific and different choice for Passover. Whipped cream is a perfect partner, with or without strawberries. For the best flavor and texture, bake the torte at least a day ahead of serving. **SERVES 10 TO 12**

1⅔ cups (170 grams) pecan halves, toasted and cooled (see page 309)

1⅓ cups (115 grams) unsweetened dried shredded coconut

6 ounces (170 grams) 55% to 72% chocolate, coarsely chopped

¾ cup (150 grams) sugar

1 cup egg whites (7 or 8 large egg whites), at room temperature

1 teaspoon pure vanilla extract

¼ teaspoon cream of tartar

½ teaspoon salt

Lightly sweetened whipped cream (see page 343; optional)

Sliced ripe strawberries (optional)

EQUIPMENT

9-by-3-inch springform pan or cheesecake pan with removable bottom

Food processor fitted with the steel blade

Stand mixer with whisk attachment

Position a rack in the lower third of the oven and preheat the oven to 350°F. Grease the pan with vegetable oil spray or butter.

In the food processor, pulse the pecans, coconut, and chocolate with ¼ cup (50 grams) of the sugar until the pecans and chocolate look very finely chopped but not completely pulverized. Set aside.

In the bowl of the stand mixer, combine the egg whites, vanilla, and cream of tartar. Beat with the whisk attachment on medium-high speed until the egg whites are creamy white and hold a soft shape when the beaters are lifted. Sprinkle the remaining sugar and the salt gradually into the egg whites, beating on high speed until the egg whites are stiff but not dry. Transfer the mixture to a much larger bowl—this is essential—for easier folding. Pour half of the pecan-coconut mixture over the egg whites. Using a large spatula, fold until the mixtures are partially blended. Pour the rest of the pecan-coconut mixture into the bowl and fold just until blended. Scrape the batter into the prepared pan and spread it evenly.

Bake for 25 to 30 minutes, or until the surface of the torte is golden brown and springs back when you press it gently with your fingers. Cool on a rack.

Slide a slim knife or spatula around the sides of the torte to detach it from the pan and remove the sides of the pan. Slide a knife between the cake and the bottom of the pan to detach it. Use a metal pancake turner to transfer the cake to a serving plate. Slice and serve plain or with dollops of lightly sweetened whipped cream and strawberries, if desired. The torte keeps, wrapped or set under a cake dome, for several days at room temperature.

COCONUT MARJOLAINE

Coconut lovers will appreciate this riff on the French classic layer cake made with coconut and rice flour instead of the traditional almonds, hazelnuts, and wheat flour. The filling is a light whipped chocolate ganache with bittersweet ganache frosting. A great professional trick for cutting and handling thin, fragile cake layers is to leave them attached to their parchment liners until the very last minute. Try it; it works. **SERVES 10 TO 12**

FOR THE COCONUT MERINGUE LAYERS

2¼ cups (190 grams) unsweetened dried shredded coconut

3 tablespoons (30 grams) white rice flour
NOTE: This recipe is not successful with Thai white rice flour.

Generous ⅛ teaspoon salt

1 cup (200 grams) sugar

9 large egg whites, at room temperature

¾ teaspoon cream of tartar

FOR THE FILLING AND FROSTING

1½ cups plus ⅔ cup heavy cream

12 ounces (340 grams) 55% to 60% chocolate, finely chopped (see Note)

Pinch of salt (optional)

1 tablespoon dark rum

EQUIPMENT

Stand mixer with whisk attachment

16-by-12-by-1-inch half sheet pan or 11-by-17-inch jelly roll pan, lined with parchment

Handheld mixer

Position a rack in the center of the oven and preheat the oven to 325°F.

To make the meringue layers, in a medium bowl, mix the coconut, rice flour, salt, and ½ cup (100 grams) of the sugar. Set aside.

Combine the egg whites with the cream of tartar in the bowl of the stand mixer. Beat with the whisk attachment on medium speed until the egg whites are creamy white and hold a soft shape when the beaters are lifted. On high speed, gradually beat in the remaining ½ cup sugar until the egg whites are stiff but not dry. Pour the coconut mixture over the meringue and fold with a rubber spatula just until incorporated. Spread the batter evenly in the lined pan.

Bake for 25 to 30 minutes, rotating the pan from back to front after about 15 minutes, until the meringue is golden brown and springy to the touch. Set the pan on a rack to cool. (The cake may be prepared up to this point 2 days ahead; cover tightly and store at room temperature.)

To make the ganache filling and frosting, bring 1½ cups of the cream to a simmer. Take the pan off the heat and stir in the chocolate until it is completely melted and smooth. Transfer 1⅓ cups of the ganache to another bowl and stir the remaining ⅔ cup cold cream into it to make a lighter ganache. Chill the light ganache for at least 2 hours or until needed. Taste and consider adding a pinch of salt to the dark ganache; leave it at room temperature to cool and thicken.

To assemble the cake, cut around the edges of the pan to detach the meringue. Grasp the edges of the parchment liner and pull or slide the meringue onto the counter or a large cutting board. Using a sharp knife or scissors, cut the meringue (and parchment) in half

crosswise, then in thirds lengthwise to make 6 layers, each about 8 by 3¾ inches.

Flip one layer over and peel off the parchment. Set the layer right side up on a sheet of foil. Remove the chilled light ganache from the refrigerator and add the rum. Beat it briefly with the handheld mixer until it is light colored and stiff enough to hold a good shape for spreading. Spread one-fifth of the ganache (about ½ cup) evenly over the layer. Place a second layer, meringue side down, on top of the ganache and press it level. Peel off the parchment and continue alternating meringue layers and ganache (remembering to peel off the parchment each time before spreading the ganache), ending with a meringue layer. Wrap and refrigerate the cake until firm, at least 1 hour.

To finish the cake, remove it from the refrigerator. Peel the parchment off the top layer. Use a sharp serrated knife and a gentle sawing motion to trim the sides of the cake evenly. Set the cake on a baking sheet (or lazy Susan or turntable if you have one). Spread the top and sides of the cake with a very thin coat of dark ganache just to create a smooth surface. If the ganache is too stiff, warm it gently by stirring with a spatula that has been dipped in hot water and wiped dry. Frost the top and sides of the cake with smooth or swirly strokes.

Use a spatula to transfer it to a serving dish or a covered container. The cake keeps, covered in the refrigerator, for at least 3 days. Remove the cake from the refrigerator 30 to 60 minutes before serving to soften the layers and open up the flavors.

NOTE: For a slightly more bittersweet filling and frosting, use 61% to 64% chocolate and make the following changes to the recipe: Increase the first amount of cream to 1¾ cups, and reduce the second amount to ½ cup. Reduce the amount of chocolate to 10 ounces. Bring the 1¾ cups of cream to a simmer and pour it over the chocolate. Mix well and transfer ¾ cup of the mixture to another bowl and combine it with the remaining ½ cup of cream. Continue as directed.

COCONUT AND MANGO TART WITH COCONUT PASTRY CREAM

This tart has loads of coconut flavor without being too sweet or cloying. If you can't find really ripe mangoes, sautéed pineapple slices make a great alternative.

SERVES 6 TO 8

FOR THE CRUST

⅓ cup (40 grams) coconut flour

1 cup plus 2 tablespoons (100 grams) unsweetened dried shredded coconut

¼ teaspoon baking powder

½ teaspoon salt

6 tablespoons (85 grams) unsalted butter, very soft

½ cup (100 grams) sugar

1 large egg white

FOR THE PASTRY CREAM AND FRUIT

1 cup Coconut Pastry Cream (page 342)

1 large ripe mango (about 320 grams/ 12 ounces)

EQUIPMENT

9½-inch fluted tart pan with removable bottom

Rimmed baking sheet

Position the oven rack at the lowest level of the oven and preheat the oven to 350°F. Grease the pan with vegetable oil spray or butter.

To make the crust, combine the coconut flour, coconut, baking powder, salt, butter, sugar, and egg white in a medium bowl and mix until all of the ingredients are blended. Press the mixture into the bottom and up the sides of the pan, making sure that the base of the pan is completely covered and the sides are thicker than the bottom.

Set the pan on the baking sheet and bake for 18 to 20 minutes, until the crust is golden brown all over.

Set the pan on a rack to cool. After about 15 minutes, push up on the bottom of the pan to loosen the sides and prevent sticking before cooling completely. Cool the crust for at least 2 hours before filling.

Leave the tart shell in the pan for support. Spread the pastry cream in the crust. Peel the mango and cut into ⅓-inch crosswise slices. Place the slices on the pastry cream. Serve within 2 hours or refrigerate. To serve, remove the sides of the pan and transfer the tart to a platter. Leftovers will keep in an airtight container in the refrigerator for up to 4 days.

COCONUT KEY LIME TART

The iconic creamy and tangy lime pie is transformed here into a sleek tart with a chewy toasted coconut crust. This is a perfect bake-ahead recipe, as the tart tastes best on the second and third days. You may use Persian limes or Key limes for this recipe. SERVES 6 TO 8

FOR THE CRUST

⅓ cup (40 grams) coconut flour

1 cup plus 2 tablespoons (100 grams) unsweetened dried shredded coconut

¼ teaspoon baking powder

½ teaspoon salt

6 tablespoons (85 grams) unsalted butter, very soft

½ cup (100 grams) sugar

1 large egg white

FOR THE FILLING

3 large egg yolks

One 14-ounce can sweetened condensed milk

1½ teaspoons finely grated lime zest

½ cup strained fresh lime juice

1 cup heavy cream, whipped (optional)

EQUIPMENT

9½-inch fluted tart pan with removable bottom

Rimmed baking sheet

Position the oven rack at the lowest level and preheat the oven to 350°F. Grease the pan with vegetable oil spray or butter.

To make the crust, combine the coconut flour, coconut, baking powder, salt, butter, sugar, and egg white in a medium bowl and mix until all of the ingredients are blended. Press the mixture into the bottom and up the sides of the pan, making sure that the base of the pan is completely covered and the sides are thicker than the bottom. Set the pan on the baking sheet and bake for 12 to 15 minutes, until the crust is just golden at the edge.

While the crust is baking, make the filling: Whisk the egg yolks, milk, and lime zest until smooth. Just before the crust is ready, whisk the lime juice into the filling. Remove the crust from the oven, scrape the filling into the hot crust, and return it to the oven. Reduce the oven temperature to 300°F and bake the tart for 12 to 14 more minutes, until the filling doesn't jiggle when you shake the pan. Set the pan on a rack to cool. After about 15 minutes, push up on the bottom of the pan to loosen the sides and prevent sticking before cooling completely. Chill the tart for at least 2 hours before serving. To serve, remove the sides of the pan and transfer the tart to a platter. The tart keeps in an airtight container in the refrigerator for up to 4 days.

Serve slices with whipped cream, if desired.

CHOCOLATE COCONUT TART

With its silky, slightly caramelized chocolate ganache and chewy coconut crust, this tart may evoke a certain (favorite) candy bar. SERVES 6 TO 8

1½ cups heavy cream

¼ cup plus 2 tablespoons (75 grams) sugar

8 ounces (225 grams) 50% to 62% chocolate, coarsely chopped

Pinch of salt

9½-inch coconut tart crust (from Coconut and Mango Tart, page 315), baked and cooled, still in the pan

Bring the cream and sugar to a simmer in a small saucepan over low heat. Remove from the heat, add the chocolate, and whisk until smooth. Return to the heat and stir in the salt. Cook over very low heat for about 10 minutes, stirring frequently—the filling should be very hot but not bubbling.

Remove from the heat and let cool for about 15 minutes. Scrape the filling into the cooled tart crust and refrigerate until cold before serving. To serve, remove the sides of the pan and transfer the tart to a platter. The tart keeps in an airtight container in the refrigerator for up to 5 days.

VARIATION: Chocolate Hazelnut Tart

Make the tart crust substituting 1¼ cups (100 grams) hazelnut meal or flour for the dried shredded coconut. Proceed as directed.

PEACH-ALMOND COBBLER

Almond flour makes a gorgeously craggy biscuit topping that's crunchy on the surface and soft within. You can try this recipe swapping in the same amount of nectarines or plums for the peaches. SERVES 6 TO 8

FOR THE FRUIT

About 2 pounds (900 grams) firm, ripe peaches, pitted and sliced about ⅓ inch thick, to make 5 cups

2 tablespoons lemon juice

½ cup (100 grams) sugar

1 tablespoon white rice flour

FOR THE TOPPING

⅔ cup (100 grams) white rice flour (preferably superfine)
—OR—
1 cup (100 grams) Thai white rice flour

⅓ cup (30 grams) almond flour/meal

⅛ teaspoon xanthan gum

1 tablespoon granulated sugar

1½ teaspoons baking powder

¼ teaspoon salt

½ cup heavy cream

¼ cup plain yogurt (any percent fat) or slightly watered down Greek yogurt

1 tablespoon coarse sugar, such as turbinado, for sprinkling

EQUIPMENT

2-quart glass or ceramic baking dish, 2 to 3 inches deep

Stand mixer with paddle attachment

Rimmed baking sheet

Position an oven rack in the lower third of the oven and preheat the oven to 400°F.

For the fruit, combine the peaches, lemon juice, sugar, and rice flour in the baking dish. Place the dish on a rimmed baking sheet and bake for 15 minutes. Stir the fruit and bake for 15 to 20 more minutes, or until juices are bubbling at the edges.

Mix the biscuit dough topping while the fruit is baking. Combine the rice and almond flours, xanthan gum, granulated sugar, baking powder, salt, cream, and yogurt in the bowl of the stand mixer and beat with the paddle attachment for 2 minutes on low speed; the dough will be very stiff. It is important to beat the dough enough or the biscuits won't rise well; don't worry about overbeating.

When the peaches are ready, spoon dollops of biscuit dough on them (don't cover the fruit completely) and sprinkle with coarse sugar. Bake for 15 to 20 minutes, or until the top is browned and the filling is bubbling in the center. Cool for at least 20 minutes. Serve warm or at room temperature. The cobbler keeps at room temperature for up to 1 day, covered with a paper towel, or for up to 3 days covered with plastic wrap and refrigerated.

ALMOND DROP BISCUITS

These drop biscuits are gently nutty in flavor, and the almond flour gives them a terrific texture and lightness. Of course, you will want to serve them with jam, but they're also delicious with mild or tangy ripe cheeses and fresh fruit. Biscuits have the best texture and rise if the dough rests in the refrigerator for at least 2 hours before baking, either scooped or unscooped. **MAKES TWELVE 3-INCH BISCUITS**

1⅓ cups (200 grams) white rice flour (preferably superfine)

—OR—

2 cups (200 grams) Thai white rice flour

⅔ cup (60 grams) almond flour/meal

2 teaspoons sugar

¼ teaspoon xanthan gum

1 tablespoon baking powder

½ teaspoon salt

1 cup heavy cream

½ cup plain yogurt (any percent fat) or slightly watered down Greek yogurt

EQUIPMENT

Stand mixer with paddle attachment

Baking sheet, lined with parchment paper

Combine the rice and almond flours, sugar, xanthan gum, baking powder, salt, cream, and yogurt in the bowl of the stand mixer and beat with the paddle attachment for 2 minutes on low speed; the dough will be very stiff. It is important to beat the dough enough or the biscuits won't rise well; don't worry about overbeating.

Cover the dough with plastic wrap and refrigerate it for at least 2 hours or up to 2 days (see headnote).

Position a rack in the upper third of the oven and preheat the oven to 400°F.

Scoop 12 high mounds of batter and drop each 2 inches apart on the lined pan. Bake for 20 to 25 minutes, until the biscuits are browned on top and bottom. Serve immediately or cool on a rack and toast before serving.

VARIATION: Almond Drop Scones

Increase the sugar in the batter to 2 tablespoons (25 grams) and sprinkle each mound of batter with a little coarse sugar before baking.

CRUNCHY COCONUT COOKIES

Crisp *and* chewy! Serve these wafers with a dish of pineapple or mango sherbet or silky rice pudding. **MAKES 3 DOZEN 2-INCH COOKIES**

Scant ½ cup (50 grams) coconut flour

1¾ cups (150 grams) unsweetened dried shredded coconut

¼ teaspoon baking powder

½ teaspoon salt

6 tablespoons (85 grams) unsalted butter, very soft

1 cup plus 2 tablespoons (225 grams) sugar

1 teaspoon pure vanilla extract

1 large egg white

¼ cup water

EQUIPMENT

2 baking sheets, lined with parchment paper

Combine the coconut flour, coconut, baking powder, salt, butter, sugar, vanilla, egg white, and water in a large bowl and mix until all of the ingredients are well incorporated. Form the mixture into a 10-inch log 2 inches in diameter on a sheet of wax or parchment paper. Wrap the log in the paper, keeping it as cylindrical as possible. Chill for at least 2 hours and up to 3 days, or wrap airtight and freeze for up to 3 months. Thaw before using.

Position racks in the upper and lower thirds of the oven and preheat the oven to 350°F.

Use a thin serrated knife to cut the dough into slices a little less than ¼ inch thick. Place the slices 1 inch apart on the lined sheets.

Bake for 12 to 14 minutes, until the cookies are golden on the bottom and browned at the edges; rotate the baking sheets from top to bottom and front to back halfway through the baking time. Place the pans on racks or slide the liners from the pans onto racks to cool completely. Repeat with the remaining dough. The cookies will keep in an airtight container for up to 1 week.

VARIATION: Crunchy Coconut Cookies with Bittersweet Chocolate

Add 2 ounces (55 grams) finely chopped semisweet or bittersweet chocolate along with the shredded coconut.

WALNUT ALFAJORES

Alfajores are luscious sandwich cookies filled with dulce de leche or cajeta (goat's-milk caramel available in cans or squeeze bottles from better supermarkets and Hispanic groceries). Every region makes alfajores with a different type of cookie, so I never hesitate to invent my own new combinations. Here, the sweet caramelized milk balances the bitter tannins in the walnuts perfectly. If you're a fan of Nutella, you can use it to fill the cookies instead of the dulce de leche. **MAKES TWENTY 2-INCH COOKIES**

Scant ½ cup (50 grams) coconut flour

1½ cups (150 grams) walnut pieces

¼ teaspoon baking powder

½ teaspoon salt

1 cup plus 2 tablespoons (225 grams) sugar

6 tablespoons (85 grams) unsalted butter, very soft

1 teaspoon pure vanilla extract

1 large egg white

⅔ cup dulce de leche or cajeta

EQUIPMENT

Food processor fitted with the steel blade

2 baking sheets, lined with parchment paper

Combine the coconut flour, walnuts, baking powder, salt, and sugar in the bowl of the food processor. Process until the walnuts are finely ground, about 15 seconds. Add the butter, vanilla, and egg white and pulse 8 to 10 times, or until the dough comes together. Form the mixture into a 10-inch log 2 inches in diameter on a sheet of wax or parchment paper. Wrap the log in the paper, keeping it as cylindrical as possible. Chill for at least 2 hours and up to 3 days, or wrap airtight and freeze for up to 3 months. Thaw before using.

Position the oven racks in the upper and lower thirds of the oven and preheat the oven to 350°F.

Use a thin serrated knife to cut the dough into slices a little less than ¼ inch thick. Place the slices 1 inch apart on the lined sheets. Bake for 9 to 11 minutes, until the cookies are golden on the bottom and browned at the edges; rotate the baking sheets from top to bottom and front to back halfway through the baking time. Place the pans on racks or slide the liners from the pans onto racks to cool completely. Repeat with the remaining dough.

When the cookies are completely cooled, fill with dulce de leche. Turn half of the cookies upside down. Spoon dulce de leche into one corner of a resealable plastic freezer bag. Clip about ¼ inch from the corner and pipe about 1½ teaspoons onto each upside-down cookie. Cover with a right-side-up cookie and press very gently to spread the filling toward the edges.

The cookies will keep in an airtight container for up to 3 days, although they will soften after the first day. Unfilled cookies may be stored for up to 1 week.

CRISPY COCONUT WAFERS

Full disclosure: I love coconut and coconut cookies and simply cannot choose a favorite between these and the Crunchy Coconut Cookies on page 321. There are only five ingredients and not a bit of flour in these easy-peasy, super-crunchy cookies. Keep them on hand to dunk, smear with peanut butter, make ice cream sammies, or dip in dark chocolate. **MAKES ABOUT 3 DOZEN 3½-INCH COOKIES**

---

3 tablespoons (45 grams) unsalted butter, melted, plus more for greasing the foil

3 large egg whites

1 cup plus 2 tablespoons (100 grams) unsweetened dried shredded coconut

½ cup (100 grams) sugar

¼ teaspoon salt

EQUIPMENT

Baking sheets

NOTES: Nonstick foil is the easier type to use and can be wiped and reused over and over again.

You can sprinkle the cookies before baking with a pinch of black or white sesame seeds, or substitute ½ cup (55 grams) finely chopped pecans or almonds with ½ teaspoon pure almond extract for an equal amount of the coconut.

Line the baking sheets with regular foil (dull side facing up) and grease the foil lightly but thoroughly with vegetable oil spray or butter. Or line the baking sheets with nonstick foil (nonstick side up, see Notes).

Mix the egg whites with the coconut, sugar, and salt until well blended. Stir in the butter. Let the batter rest for at least 15 minutes to allow the coconut to absorb moisture (or cover and store it in the refrigerator for up to 3 days).

Position racks in the upper and lower thirds of the oven and preheat the oven to 300°F.

Stir the batter well. Drop level tablespoons 3 inches apart on the lined sheets. Spread the batter to a diameter of 3½ inches (about ¼ inch thick).

Bake for 20 to 25 minutes, or until the cookies are mostly golden brown on top—a few pale patches are okay. Rotate the sheets from front to back and top to bottom about halfway through the baking time. If the cookies are not baked enough, they will not be completely crisp when cool.

Slide the foil sheets onto racks and let the cookies cool slightly or even completely before removing them.

To retain crispness, put the cookies in an airtight container as soon as they are cool. They may be stored airtight for at least 1 month.

CHUNKY DOUBLE-CHOCOLATE COCONUT MERINGUES

Add coconut flakes, salted almonds, and chunks of bittersweet and creamy coconut-flavored white chocolate to a light meringue cookie, and you get a riot of creamy, crunchy, chewy, sweet, and salty in every bite. **MAKES 45 TO 50 COOKIES**

---

1 cup (140 grams) roasted salted almonds, coarsely chopped

4 ounces (115 grams) 70% chocolate, cut into chunks, or ⅔ cup purchased chocolate chunks or chips

1 cup (40 grams) unsweetened dried flaked coconut (coconut chips)

2 ounces (60 grams) coconut white chocolate (such as Lindt), cut into ⅓-inch squares

3 large egg whites, at room temperature

⅛ teaspoon cream of tartar

½ cup plus 2 tablespoons (125 grams) sugar

EQUIPMENT

Stand mixer with whisk attachment, or handheld mixer

2 baking sheets, lined with parchment paper

Position racks in the upper and lower thirds of the oven and preheat the oven to 200°F.

In a small bowl, combine one-quarter of the almonds, dark chocolate, and coconut for sprinkling. Set aside.

In a medium bowl, mix the remaining almonds, chocolate, and coconut with the coconut white chocolate. Set aside.

Combine the egg whites and cream of tartar in the bowl of the stand mixer (or in a large bowl if using a handheld mixer). Beat on medium-high speed (or on high speed with the handheld mixer) until the egg whites are creamy white and hold a soft shape when the beaters are lifted. Continue to beat on medium to high speed, adding the sugar a little at a time, for 1½ to 2 minutes, until the egg whites are very stiff and have a dull sheen. Use a large rubber spatula to fold in the mixture of nuts, coconut, and both chocolates, just until blended.

Drop generous tablespoons of meringue 1½ inches apart on the lined baking sheets. Make sure all of the batter fits on the two sheets so all can be baked at once; if necessary, make each cookie a little bigger. Sprinkle the meringues with the reserved chocolate, almonds, and coconut.

Bake for 1½ hours, rotating the pans from top to bottom and from front to back halfway through the baking time to ensure even baking. Remove a "test" meringue and let it cool completely before taking a bite (meringues are never crisp when hot). If the test meringue is completely dry and crisp, turn off the oven and let the remaining meringues cool completely in the oven. If the test meringue is soft or chewy or sticks to your teeth, bake for another 15 to 30 minutes before cooling in the oven.

To prevent cookies from becoming sticky, put them in an airtight container as soon as they are cool. The cookies keep for at least 2 weeks.

COCONUT MERINGUES

Meringues always make great (and long-keeping) cookies that can also be piled into dessert glasses with whipped cream and fruits—think pineapple, mango, or bananas to go with the coconut—for an instant seasonal dessert. You can sandwich these with ice cream and serve them with chocolate sauce. **MAKES 30 TO 40 COOKIES**

⅓ cup (30 grams) unsweetened dried shredded coconut

¾ cup (75 grams) blanched or unblanched almond flour/meal (optional)

1 cup (200 grams) sugar

4 large egg whites, at room temperature

¼ teaspoon cream of tartar

EQUIPMENT

Baking sheets, lined with parchment paper

Stand mixer with whisk attachment, or handheld mixer

Position racks in the upper and lower thirds of the oven and preheat the oven to 200°F.

In a small bowl, whisk the coconut (and almond flour, if using) with ⅓ cup of the sugar. Set aside.

Combine the egg whites and cream of tartar in the bowl of the stand mixer fitted with the whisk attachment (or in a large bowl if using a handheld mixer). Beat on medium-high speed (or on high speed with the handheld mixer) until the egg whites are creamy white and hold a soft shape when the beaters are lifted. Continue to beat, adding the remaining sugar a little at a time, for 1½ to 2 minutes, until the egg whites are very stiff and have a dull sheen. Pour the coconut mixture over the egg whites and fold in with a rubber spatula just until combined.

Drop heaping tablespoons of meringue, or any size and shape you like, 1½ inches apart onto the lined baking sheets.

Bake for 1½ hours. Rotate the pans from top to bottom and from front to back halfway through the baking time to ensure even baking. Remove a "test" meringue and let it cool completely before taking a bite. (Meringues are never crisp when hot.) If the meringue is completely dry and crisp, turn off the heat and let the remaining meringues cool completely in the oven. If the test meringue is soft or chewy or sticks to your teeth, bake for another 15 to 30 minutes before testing again.

To prevent the meringues from becoming moist and sticky, put them in an airtight container as soon as they are cool. They may be stored airtight for 2 weeks.

BROWN SUGAR PECAN COOKIES

These plain, dense, seemingly rustic cookies will sneak up on you. They are adapted from a recipe in Rosetta Costantino's compelling book *Southern Italian Desserts*. Authentic *dolci di noci*, made with walnuts and white sugar and without salt, are quintessentially Italian, and very, very good. I gave them an American flavor twist with pecans instead of walnuts, a bit of brown sugar, and some salt. Don't miss the peanut variation with vanilla. If you want to gild the lily, bury a dried strawberry in each cookie before baking. Or circle back to the recipe's Italian roots—use walnuts instead of pecans and hide Amarena cherries in the cookies.

MAKES THIRTY-TWO 1½-INCH COOKIES

2¾ cups (285 grams) pecan halves or large pieces

½ cup (100 grams) granulated sugar

½ cup (100 grams) packed light or dark brown sugar

¼ teaspoon salt

1 large egg

EQUIPMENT

Food processor fitted with the steel blade

Baking sheet, lined with parchment paper

Position a rack in the center of the oven and preheat the oven to 350°F.

Combine the nuts, both sugars, and salt in the food processor. Process, pulsing frequently, until the nuts are finely ground. Add the egg and process to a coarse, sticky dough.

Remove the dough and knead it with your hands a few times to be sure it is evenly mixed. On a cutting board or sheet of wax or parchment paper, pat the dough into a 6-by-4-inch rectangle about 1 inch thick. Use a straightedge or metal spatula to square the sides. Cut the rectangle lengthwise into 4 strips and crosswise into 8 to make 32 equal pieces. For the quickest option, space the pieces 1 inch apart on the lined sheet. Or roll each piece into a ball or a log shape.

Bake for 12 to 15 minutes, or until the surface of the cookies is partially browned and the bottoms are deep brown. Set the baking sheet on a rack to cool. Store the cooled cookies in an airtight container.

VARIATION: Nutty Peanut Cookies

Substitute 2¼ cups (9 ounces/255 grams) roasted unsalted peanuts for the pecans. Increase the salt to ⅜ teaspoon and add 1 teaspoon pure vanilla extract with the egg.

PEANUT CRUNCH BROWNIES

This very rich, very decadent brownie with a not-quite-grown-up flavor appeals to most of the grown-ups I know. **MAKES 25 SMALL BROWNIES**

FOR THE CHOCOLATE LAYER

8 tablespoons (1 stick/115 grams) unsalted butter, softened

4 ounces (115 grams) 66% to 72% chocolate, coarsely chopped

1¼ cups (250 grams) granulated sugar

2 large eggs

1 teaspoon pure vanilla extract

¼ teaspoon salt

⅓ cup plus 1 tablespoon (60 grams) white rice flour
—OR—
½ cup plus 1 tablespoon (60 grams) Thai white rice flour

FOR THE PEANUT CRUNCH LAYER

4 tablespoons (½ stick/55 grams) unsalted butter, softened

½ cup (100 grams) packed light brown sugar

¼ teaspoon salt

1 large egg

½ teaspoon pure vanilla extract

½ cup plus 2 tablespoons (160 grams) salted crunchy natural peanut butter

⅓ cup plus 1 tablespoon (60 grams) white rice flour
—OR—
½ cup plus 1 tablespoon (60 grams) Thai white rice flour

¼ teaspoon baking soda

EQUIPMENT

9-inch square metal baking pan

Position a rack in the upper third of the oven and preheat the oven to 350°F. Line the baking pan across the bottom and up all four sides with foil.

For the chocolate layer, place the butter and chocolate in a medium stainless steel bowl set directly in a wide skillet of barely simmering water. Stir frequently until the chocolate is completely melted and the mixture is smooth and quite hot to the touch. Stir in the sugar and remove from the heat. Use a large spoon to beat in the eggs, vanilla, and salt. The mixture should be very smooth; if it is not, place the bowl back in the water bath over very low heat for half a minute, stirring constantly. Add the flour and beat with a spoon until the batter comes away from the sides of the bowl, about 1 minute. Set aside.

In a separate bowl, use a large spoon or spatula to mix all of the ingredients for the peanut crunch layer until completely blended. Press the mixture evenly in the bottom of the prepared pan. Bake for 8 to 10 minutes, until it is slightly darker at the edges.

Remove the pan from the oven and spoon dollops of chocolate batter all over the hot crust. Spread gently to make an even layer. Return the pan to the oven and bake for 20 to 22 minutes, until the surface is dry and pulls away slightly from the sides of the pan. Cool completely in the pan on a rack. Lift the edges of the pan liner and transfer the brownies to a cutting board. Cut into 25 squares. The brownies keep for 3 to 4 days in an airtight container.

VARIATION: Hazelnut Crunch Brownies

Substitute 2 cups (160 grams) hazelnut meal/flour plus ½ teaspoon salt for the peanut butter.

CHAPTER NINE

ELEMENTS

This chapter is a helpful little pantry of items, including fillings, frostings, toppings, and crunchy elements that are integral to several of the recipes. They can be used to customize or accompany the recipes.

TWO CUPS
16 oz
1 3/4
1 2/3
1 1/2
1 1/3
1 1/4
ONE CUP
8 oz
3/4
2/3
1/2
1/4 CUP

CLARIFIED BUTTER

Clarified butter is called for in several of the sponge cakes (génoises) because it has no water that might interfere with the structure of the cake. Sometimes referred to as drawn butter or ghee, clarified butter is pure melted butter minus the milk solids (foam that usually accumulates on top) and the water that sinks to the bottom. You can make it in a saucepan, but it's easier, and you will get a higher yield, in the microwave. Store any leftover clarified butter in the fridge or freezer and use it to fry or sauté; you can heat it to a higher temperature than regular butter without burning.

MAKES ABOUT 6 TABLESPOONS (85 GRAMS)

---

8 tablespoons (1 stick/115 grams) unsalted butter

In a microwave: Use full power in a 600- to 800-watt microwave or medium power in a larger-watt oven. In any case, keep watch. Cut the butter into 1-tablespoon chunks (for even melting with minimal spitting) and put it in a heatproof jar or Pyrex measuring cup that is taller than it is wide; a 1-cup Pyrex measure or similar size jar is perfect for 1 stick of butter. In any case, choose a container tall enough that the butter can rise without overflowing.

Heat the butter until it is melted and rising. You have to watch it through the window: let it rise toward the top of the container, then stop the oven. If the butter is not separated almost perfectly with foam on top, clear yellow clarified butter in the middle, and a little watery liquid on the bottom, let it settle down again, then microwave and let it rise once more (watching carefully). The second rise should give you almost perfectly separated layers, but you can give it a third rise if you are not satisfied.

Remove the butter from the oven and let it sit undisturbed for a few minutes (or longer, but not long enough to resolidify). Tilt the container and spoon off and discard the layer of foam. Pour and/or spoon the clear clarified butter into another container, leaving the watery liquid behind.

On the stove: Heat the butter until sizzling hot in a small saucepan. Wait for a minute until the foam settles on top. Tilt the pan and skim off and discard the foam. Carefully spoon the clear yellow oil (clarified butter) into a clean dish, leaving behind the watery (actually, it looks kind of milky) mixture at the bottom of the pot.

PRUNES POACHED IN COFFEE AND BRANDY

This sweet velvety fruit compote is particularly good with Buckwheat Sponge Cake (page 171), Oat Flour Sponge Cake (page 86), and the brown rice sponge cake used to make Brown Rice Sponge Cake with Three Milks (page 57), and is a great way to enjoy the flavor of those cakes simply, with whipped cream rather than sweeter fillings and frostings. Any leftovers should be eaten with vanilla ice cream! **MAKES ABOUT 4 CUPS**

---

2 cups strong regular coffee or decaf

1 cup (200 grams) sugar

1 teaspoon pure vanilla extract

2 tablespoons brandy

4 cups (750 grams) pitted prunes (dried plums)

Combine all of the ingredients in a medium saucepan and bring the mixture to a simmer. Simmer for 5 to 10 minutes, turning the prunes gently once or twice, or until they are just tender. Serve hot, warm, or chilled. Poached prunes keep in the refrigerator for at least a week.

WALNUT OR PECAN PRALINE BRITTLE

Chopped, crushed, or pulverized, caramelized nuts ("praline" in the world of French pastry) are a secret weapon for adding flavor and crunch to all kinds of plain or fancy desserts: use them as a garnish for creamy or soft desserts, or fold them into ice creams, buttercream, whipped cream, or pastry cream. Praline is versatile, flavor intense, and simple to make. You can substitute any nut or make a half recipe if you like. Do not use a nonstick pan; the melted sugar will ruin it.

MAKES ABOUT 2⅔ CUPS FINELY CHOPPED PRALINE

---

1 cup water

¼ teaspoon cream of tartar

2 cups (400 grams) sugar

2⅔ cups (175 grams) coarsely chopped walnuts or pecans

EQUIPMENT

Baking sheet, lined with foil or parchment paper

Have the lined baking sheet ready near the stove, along with a skewer or a very small knife and a small white plate to test the color of the syrup. Pour the water into a 2- or 3-quart saucepan over medium heat and sprinkle in the cream of tartar. Add the sugar, pouring it in a thin stream into the center of the pot to form a low mound. Don't stir, but if necessary use your fingers to pat the sugar mound down until it is entirely moistened. Any sugar touching the edges of the pan should be below the waterline.

Cover the pot and cook for a few minutes, without stirring, until the sugar is dissolved. Uncover the pan and cook without stirring until the syrup begins to color slightly. Swirl the pot gently rather than stirring if the syrup seems to be coloring unevenly.

Use the skewer or the tip of the knife to drop a bead of syrup on the plate from time to time. When a drop looks pale amber, add the nuts and turn them gently with a clean, dry wooden spoon or silicone spatula just until they are completely coated with syrup. (Brisk stirring will cause the caramel to crystallize.) Continue to cook, pushing the nuts around gently if the syrup is coloring unevenly, until a drop of syrup looks deep amber on the plate. Immediately scrape the mixture onto the lined baking sheet and spread it out as well as you can.

When the praline is still quite warm, use dry hands to break it into pieces and transfer it (still warm) on its parchment to a zipper-lock bag or airtight container; seal the bag immediately to prevent the praline from getting sticky. Store airtight for up to a month. When ready to use, break the praline into smaller pieces. To make praline powder, chop or pulse the pieces in a food processor.

CARDAMOM BRITTLE

Nibble these caramel shards like you would exotic candy, or serve them intact or crushed with a creamy dessert such as Silky Saffron Rice Pudding (page 78) or Silky Chocolate Pudding (page 80), with any ice cream, or with Almond Butter Cake (page 300) and sliced oranges. They're sweet and spicy and bitter all at the same time. You can slip them into a cellophane bag tied with a ribbon and bring them to your favorite hostess. The almonds get deeply toasted and almost burnt, in a good way. For a lighter statement, add the almonds when the caramel is close to done. Do not use a nonstick pan; the melted sugar will ruin it. **MAKES ABOUT 1 POUND OF CANDY**

---

1 cup water

¼ teaspoon cream of tartar

2 cups (400 grams) sugar

½ cup (45 grams) raw sliced almonds

1 teaspoon cardamom seeds (from about 12 pods)

1 tablespoon finely grated orange zest

½ teaspoon coarse salt (optional)

EQUIPMENT

Baking sheet, lined with parchment paper

Set the lined baking sheet near the stove along with a skewer or small knife, a silicone spatula or wooden spoon, and a small white plate. Have all of the ingredients measured and at hand as well.

Pour the water into a 2- or 3-quart saucepan and set over medium heat. Sprinkle in the cream of tartar, then add the sugar, pouring it in a thin stream into the center of the pan to form a low mound. Don't stir, but if necessary, use your fingers to pat the sugar down until it is entirely moistened. Any sugar touching the edges of the pan should be below the waterline.

Cover the pan and cook, without stirring, for a few minutes until the sugar is dissolved. Uncover the pan and cook without stirring until the syrup begins to color slightly.

Use the skewer or the tip of the knife to drop a bead of syrup onto the plate from time to time. When a drop of caramel looks pale amber, add the almonds to the pan and turn them gently with the silicone spatula or wooden spoon just until thoroughly coated (brisk stirring can cause the caramel to crystallize). Continue to cook, pushing the nuts around gently if the caramel is cooking unevenly, until a drop of caramel looks deep amber on the plate. Remove the pan from the heat and immediately stir in the cardamom seeds and orange zest until well mixed. Scrape the mixture onto the lined pan and spread it as thin as you can. Sprinkle with the salt, if desired.

When the brittle is cool enough to handle, but still quite warm, use dry hands to break it into long shards and slide them (still quite warm) on the parchment paper into a zipper-lock bag or other airtight container. Seal the bag immediately to prevent the shards from getting sticky. Brittle keeps at room temperature in an airtight container for at least a month.

VARIATION: Cinnamon-Orange Brittle

Substitute 1 teaspoon ground cinnamon for the cardamom seeds.

LEMON OR LIME CURD

Every dessert maker needs a recipe for lemon curd. This one is easy and versatile (you can use whole eggs or just yolks or a combination). Use it to fill the Lemon Tart (page 154) or the Lemon Cream Roulade with Strawberry-Mint Salad (page 60). Leftovers make a luxurious spread for toast or biscuits, or a quick last-minute sandwich cookie or thumbprint cookie filling. The recipe makes more than enough to fill a 9½-inch tart. Make a half recipe, or even one-third recipe for smaller projects. **MAKES 3 CUPS (ENOUGH FOR A 9½-INCH TART WITH SOME LEFT OVER)**

---

6 large eggs, or 2 large eggs plus 6 egg yolks

Grated zest of 2 medium lemons or 3 limes

1 cup strained fresh lemon or lime juice (from about 6 juicy medium lemons or 10 to 12 limes)

1 cup (200 grams) sugar

12 tablespoons (1½ sticks/170 grams) unsalted butter, cut into chunks

EQUIPMENT

Medium-mesh strainer

Set the strainer over a medium bowl near the stove. Whisk the eggs in a medium nonreactive saucepan to blend. Whisk in the lemon zest, juice, and sugar. Add the butter chunks. Whisk over medium heat, reaching into the corners and scraping the sides and bottom of the pan, until the butter is melted and the mixture is thickened and beginning to simmer around the edges, then continue to whisk for about 10 seconds longer. Remove from the heat and scrape into the strainer, pressing gently on the solids. Scrape any lemon curd clinging to the underside of the strainer into the bowl. Let cool. The curd keeps in a covered container in the refrigerator for up to a week.

THE NEW VANILLA PASTRY CREAM

This rice-flour-based pastry cream is the creamy, custardy liaison between fresh fruit and buttery crisp or chewy tart crusts. You'll use it in the Strawberry-Blueberry Tart (page 275), the Coconut and Mango Tart (page 315), and New Classic Boston Cream Pie (page 145). Taste and compare it to your usual pastry cream—made with wheat flour and/or cornstarch—and you may make it your new basic.

Super smooth and silky, this pastry cream is easier to make than regular pastry cream. Use superfine or Thai white rice flour for this. If you like lots of custard in your tarts, double the recipe. **MAKES ABOUT 1 CUP**

---

3 tablespoons (40 grams) sugar

1 tablespoon plus 2 teaspoons (13 grams) superfine white rice flour (from Authentic Foods; see Resources, page 351)
—OR—
1 tablespoon plus 2 teaspoons (13 grams) Thai white rice flour

1 cup milk

2 large egg yolks

¾ teaspoon pure vanilla extract

EQUIPMENT

Medium-fine-mesh strainer

Set the strainer over a bowl near the stove. Whisk the sugar and rice flour together in a small heavy-bottomed saucepan. Whisk in about 2 tablespoons of the milk to make a smooth paste. Whisk in the egg yolks until smooth; whisk in the rest of the milk. Cook over medium heat, stirring constantly, sweeping the bottom, sides, and corners of the pan to prevent the mixture from scorching.

When the mixture begins to simmer, set a timer for 5 minutes and continue to cook and stir, turning down the heat if necessary to barely maintain a simmer.

Immediately scrape the custard into the strainer. Stir the custard through it, but don't press on any bits of cooked egg that may be left behind. Scrape the custard clinging to the underside of the strainer into the bowl as well. Stir in the vanilla extract. Let cool for about half an hour, then cover with wax paper or plastic wrap pressed directly against the surface of the custard to prevent a skin from forming. Refrigerate until chilled. (The pastry cream can be refrigerated for up to 3 days.)

VARIATIONS

Lemon Custard Pastry Cream

Add 1 teaspoon lightly packed grated lemon zest with the rice flour. Omit the vanilla extract.

Coconut Pastry Cream

Substitute 1 cup full-fat Asian coconut milk (not the beverage) for the milk. Be careful not to boil too vigorously, or it may separate.

WHIPPED CREAM

Whipped cream is one of the best and most versatile of dessert toppings and fillings. Unsweetened or just lightly sweetened (with or without vanilla), whipped cream adds a cool, creamy counterpoint to the main flavors and texture of a dessert in a way that often makes those flavors pop. Adding a little yogurt (see Tangy Whipped Cream, page 345) lends depth and interest to the cream—adding more yogurt, sour cream, or crème fraîche transforms it further. And then, of course, there are many different ways to flavor it. **MAKES 2 TO 2½ CUPS**

---

1 cup heavy cream

½ teaspoon pure vanilla extract (optional)

2 teaspoons sugar, or more to taste (optional)

EQUIPMENT

Handheld mixer

NOTE: Whipped cream used to frost or fill a cake should always be underwhipped because spreading the cream over the cake causes it to become stiffer. If you begin frosting with stiffly beaten cream, you will end up with overbeaten cream.

Using chilled beaters (or a whisk), beat the cream with the vanilla and sugar (if using) in a chilled bowl, until it holds a very soft shape. Taste and adjust the sugar. Beat to the desired consistency. Whipped cream is ideal if used on the day it's made, but it does keep in the refrigerator for 2 or 3 days, or up to 5 days if it is mixed with a little yogurt (see the Tangy Whipped Cream variation). Rewhisk the cream briefly to reincorporate any liquid that may have settled to the bottom of the bowl.

(recipe continues)

VARIATIONS

Tangy Whipped Cream

Whip the cream with or without vanilla and with 1 tablespoon sugar, as directed, until nearly as stiff as you want it to be. Add 1 tablespoon plain yogurt (any fat content) and whisk just until combined.

Coffee Whipped Cream

Add 2 teaspoons instant espresso powder or 2½ teaspoons freeze-dried instant coffee crystals to the cream and increase the sugar to a generous tablespoon. Beat as directed, tasting and adjusting the sugar toward the end.

Rose Whipped Cream

Add ½ teaspoon rose water with the sugar and beat as directed, tasting and adjusting the sugar toward the end.

Whipped Crème Fraîche

Substitute chilled crème fraîche for any or all of the cream.

Honey Whipped Cream

Beat the cream with a generous tablespoon of honey and omit the sugar at the beginning. Beat as directed, correcting the sweetness with sugar toward the end.

Praline Whipped Cream

Make whipped cream as directed, but without any sugar. Fold in any amount of finely chopped praline or praline powder (see page 337), from ⅓ cup to 1⅓ cups, to your taste.

MOCHA MOUSSE FROSTING

Coffee balances the sweetness of the chocolate in this elegant and very flavorful frosting with a light creamy texture. Use it in lieu of the Milk Chocolate Frosting when you make the chocolate fudge cake on page 103. **MAKES 1 QUART FROSTING, ENOUGH FOR ONE 8- OR 9-INCH CAKE WITH 2 OR 3 LAYERS OR 24 CUPCAKES**

½ cup strong regular coffee or decaf (hot or warm)

8½ to 9 ounces (240 to 255 grams) milk chocolate, coarsely chopped

1½ cups heavy cream

Put the coffee and chocolate in a medium stainless steel bowl. Bring an inch of water to a simmer in a wide skillet. Turn off the heat and set the bowl of chocolate in the water. Let it rest for 15 minutes, gently shaking the bowl several times to submerge the chocolate. When the chocolate is melted, start whisking at one edge and continue whisking until all of the chocolate is incorporated and the mixture is smooth.

When ready to frost a cake, have the layers unmolded and ready to assemble. Make sure the chocolate mixture is just warm enough to pour. Whip the cream until firm peaks form. Gently whisk about a third of the whipped cream into the chocolate mixture, then scrape it into the whipped cream and gently whisk until uniform in color. Working quickly before it sets up, transfer about a third of the mousse to the first cake layer, spread it evenly, and top with the other cake layer. Press down gently to level the cake. Skip the crumb-coat step; scrape the remaining mousse immediately onto the cake; spread over the sides and smooth the top. Do not overwork the mousse or it will become grainy.

SARAH BERNHARDT CHOCOLATE GLAZE

This is the perfect bittersweet chocolate glaze for cakes and desserts that are kept in the refrigerator because it does not crack when chilled. Use it for Triple Chocolate Layer Cake (page 234) and Chocolate Raspberry Celebration Cake (page 240). **MAKES ABOUT 1⅔ CUPS**

---

8 ounces (225 grams) 55% to 70% chocolate (see Note), coarsely chopped

12 tablespoons (1½ sticks/170 grams) unsalted butter, cut into pieces

1 tablespoon light corn syrup

1 tablespoon plus 2 teaspoons water

EQUIPMENT

Instant-read thermometer

NOTE: The higher the percentage of cacao in the chocolate, the stronger and more bitter the glaze.

Place the chocolate, butter, corn syrup, and water in a heatproof bowl set in a wide skillet of barely simmering water. Stir frequently and gently (to prevent air bubbles) with a spatula or wooden spoon until almost completely melted. Remove the glaze from the water and set aside to finish melting, stirring once or twice until the glaze is perfectly smooth. Or melt in a microwave on Medium (50% power) for about 2 minutes. Stir the mixture gently until completely smooth; do not whisk or beat.

Cool the glaze to 90°F before pouring it over a cake. If you want to crumb-coat the cake first, let the glaze cool without stirring, until it is nearly set and the consistency of easily spreadable frosting; then spread a very thin coat of glaze over the cake to smooth the surface and glue on any loose crumbs. Reheat the remaining glaze gently to 90°F (for best shine) before pouring it over the cake.

NEW CLASSIC BUTTERCREAM

This is real French buttercream reorganized to eliminate the trickiest steps. Traditional French buttercream requires pouring hot sugar syrup over eggs while beating steadily—without scrambling them or splattering most of the syrup around the sides of the bowl. Then the mixture is reheated (to be sure that the eggs get cooked but not scrambled) and beaten again to cool before beating in the butter. Pastry chefs make this all the time, but home cooks are at a disadvantage because small batches are trickier than large batches. This recipe produces classic results with fewer, easier steps. **MAKES ABOUT 3 CUPS, MORE THAN ENOUGH FOR ONE 8- OR 9-INCH LAYER CAKE**

---

4 large egg yolks or 2 whole large eggs, at room temperature

Large pinch of salt

¼ cup water

⅔ cup (130 grams) sugar

¾ pound (3 sticks/340 grams) unsalted butter, slightly softened but not too squishy

EQUIPMENT

Stand mixer with paddle attachment

Medium-fine strainer

Instant-read thermometer

Set a strainer over the bowl of the stand mixer.

In a medium stainless steel bowl (because glass bowls are too slow to heat), whisk the egg yolks (or whole eggs), salt, and water together thoroughly. Whisk in the sugar.

Set the bowl in a wide skillet filled with enough hot water to reach above the depth of the egg mixture. Over medium heat, with a heatproof silicone spatula, stir the egg mixture, sweeping the sides and bottom of the bowl constantly to prevent the eggs from scrambling. Adjust the burner so the water barely simmers and continue to stir until the mixture registers between 175°F and 180°F on an instant-read thermometer. Swish the thermometer stem in the hot skillet water to rinse off the raw egg after reading the temperature each time.

Remove the bowl from the skillet and scrape the mixture into the strainer. Rap the strainer to coax the mixture through it, but do not press on any bits that are left in the strainer. Turn the strainer and scrape the mixture clinging to the underside into the bowl of the stand mixer. Beat with the paddle attachment on high speed for 3 to 5 minutes, or until the mixture is cool and resembles a fluffy, foam-like soft whipped cream.

Beat the butter into the foam a tablespoon at a time. The foam will deflate as you add the butter. If the butter is a little too cold, the mixture may curdle or separate at first, but it will smooth out as you continue to beat, and if it doesn't you can set it in the warm water in the skillet for a few seconds and then continue to beat.

If the foam is still warm or the butter too soft when you combine them, the buttercream will seem a little soupy instead of thick and creamy. If it doesn't come together and thicken with beating, set the bowl in a bowl of ice and water or in the refrigerator for 5 to 10 minutes; then resume beating until the mixture is creamy and smooth.

Use the buttercream right away or refrigerate it until needed. Buttercream keeps in a sealed container in the refrigerator for several days or in the freezer for 3 months. To soften, break chilled or frozen buttercream into chunks with a fork. Microwave on Low for just a few seconds, then stir with a rubber spatula. Or set the bowl in hot water until some buttercream melts around the sides of the bowl. Remove the bowl and stir. If the buttercream is not smooth and spreadable, repeat the gentle warming and stirring steps until it is.

Tips for Working with Buttercream

Buttercream can be slightly temperamental. Keep it smooth as you work with it by stirring it briskly with a rubber spatula from time to time—do this anytime it begins to look curdled or separated or even full of air bubbles.

Every time you refrigerate or chill buttercream, you will have to bring it back to working consistency by stirring briskly (if it is not yet too hard) or by very gently warming it (as described above) to soften it slightly before you stir it. If you accidentally oversoften it, it will turn to soup. Simply chill it again as described and carry on.

RESOURCES

INGREDIENTS

Allen Creek Farm

www.chestnutsonline.com
The best source for fresh, flavorful, extra-fine chestnut flour without any smoky flavor. They also make a coarser stone-ground chestnut flour. The recipes in this book use extra-fine flour.

Anson Mills

www.ansonmills.com
Rustic aromatic buckwheat flour.

Arrowhead Mills

www.arrowheadmills.com
Organic white and brown rice flours. Regular grind rather than superfine. These flours may give less light and delicate results than superfine flours in some recipes, but they are completely acceptable unless otherwise indicated in the recipe.

Authentic Foods

www.authenticfoods.com
A great source for superfine flours: white and brown rice, white corn, sorghum, and more.

Bob's Red Mill

www.bobsredmill.com
Bob's Red Mill makes every kind of flour imaginable, including corn flour, oat flour, white and brown rice flour, coconut flour, sorghum flour, teff flour, buckwheat flour, and xanthan gum. Their rice, sorghum, and corn flours are regular grind and may therefore give less light and delicate results than superfine flours in some recipes; however, they are completely acceptable unless otherwise indicated in the recipe.

King Arthur Flour

www.kingarthurflour.com
Cake and tart pans, kitchen utensils, scales, digital scales, parchment paper sheets, flours, xanthan gum, specialty sugars, and more.

Market Hall Foods

www.markethallfoods.com
A favorite purveyor of specialty foods and ingredients, with retail stores in Oakland and Berkeley, California. Market Hall has a subspecialty in baking ingredients, including chocolates and cocoa, honeys, nuts, pastes, specialty sugars, preserves, a variety of chestnut products, and much more.

Parrish's Cake Decorating Supply, Inc.

225 West 146th Street
Gardena, CA 90248
800-736-8443
They sell my favorite Magic Line cake pans, which are available in myriad sizes with and without removable bottoms. Shop at the store or by catalogue or phone. Wasserstrom (www.wasserstrom.com) also carries some Parrish cake pans.

Penzeys Spices

www.penzeys.com
This impeccable spice source sells every kind of herb, spice, and seasoning imaginable. The catalogue alone is an education in flavor ingredients and their uses.

Sur La Table

www.surlatable.com
Purveyor of all kinds of baking supplies and utensils, including digital scales, Magic Line pans, and fluted tart pans with removable bottoms, as well as scales.

The Teff Company

www.teffco.com
Brown and ivory teff grains and flour, plus recipes, history, and information about teff.

LOCAL SHOPPING

Health food and natural food stores and high-end grocery stores carry special flours, nuts, and seeds in packages or in bulk. Big-box stores are also good sources.

Asian grocery stores carry the finest grind and most economical white rice flour and sweet white rice flour (also known as glutinous white rice flour) from Thailand. You can find Thai flour online at Amazon.com, efooddepot.com, or asianfoodssuperstore.com. Look for the Erawan brand (marketed by Erawan Marketing in Bangkok) that was used in testing for this book or for the Flying Horse brand. Note that Thai rice flour bags are color-coded: red printing on the label for regular white rice flour; green printing for sweet rice flour.

Restaurant supply stores are treasure troves for reasonably priced pans and utensils.

GLUTEN-FREE LIVING

The following links provide information about FDA regulations for gluten-free product and ingredient labeling, gluten-free certification, and more.

www.celiac.org/live-gluten-free/glutenfreediet/label-reading/
www.glutenfreegigi.com/gluten-free-product-certification-how-it-works-and-what-it-means-to-you/

APPENDIX

Flour Weights and Volumes

My colleague Maya and I both measured by weight in our separate kitchens so that we always knew that we were measuring the same ingredient in the same way. To translate our weights into cups and spoons, we agreed on one method for using measuring cups (see page 29) and then weighed multiple cups of each flour to establish an average per-cup weight for each kind of flour from which to derive volume measurements.

We weighed several cups of each flour straight from new bags and from open bags or canisters that had been sitting for a while. We did not reweigh previously measured flour or put it back into the bag or canister before we had finished the whole weighing experiment. Why? Because flour gets looser and fluffier every time you handle it: one level cup of flour dumped out and remeasured will weigh less than before, because it won't all fit back in the cup! In fact, after measuring and weighing all of the flour from the bag it came in, cup by cup, it wouldn't fit back in the bag unless we shook and rapped the bag against the counter to settle and compact it.

Our average actual weights for some of the flours were consistent with the information on the back of the flour bag: that is, if the serving size on the nutritional panel was ¼ cup (30 grams), our cup of that flour actually weighed 120 grams. But our weights for other flours were considerably different from the information on the bag. After much consideration, we decided to go with our own actual weights.

It has to be said that our translation from weights to cups and tablespoons is quirky at times because we chose not to round up to the nearest ¼ or ⅓ cup if rounding up made too much difference to the true weight. We used the following weights per cup of flour to determine the volume measures for our recipes. But when all is said and done, your results will be most similar to those we got in our own kitchens if you measure all of your flours with a scale. Once you start, I guarantee you'll never go back.

Buckwheat Flour

- Buckwheat flour from Bob's Red Mill: 125 grams per cup

Chestnut Flour

- Chestnut flour from Allen Creek Farm: 100 grams per cup

Coconut Flour

- Coconut flour from Bob's Red Mill: 110 grams per cup

Corn Flours

- Corn flour from Bob's Red Mill: 125 grams per cup
- Superfine white corn flour from Authentic Foods: 140 grams per cup
- Stone-ground cornmeal: 160 grams per cup

Nut Flours/Meals

The following weights reflect average weights of purchased flours or meals. The volume measurements in the recipes in this book are based on these weights. Homemade flours can be considerably different in weight per cup, depending on how you make them (see page 355 for methods and discussion); measure homemade flour by weight rather than volume to compensate for differences.

- Almond flour/meal: 100 grams per cup
- Hazelnut flour/meal: 80 grams per cup
- Pecan flour/meal: 85 grams per cup
- Pistachio flour/meal: 100 grams per cup
- Walnut flour/meal: 95 grams per cup

Oat Flour

- Oat flour from Bob's Red Mill or Arrowhead Mills: 100 grams per cup

Rice Flours

- Superfine white rice flour from Authentic Foods or white rice flour from Bob's Red Mill: 140 to 160 grams per cup
- Thai white rice flour (Erawan): 100 grams per cup
- Glutinous (aka sweet) rice flour from Bob's Red Mill: 160 grams per cup
- Superfine glutinous (aka sweet) rice flour from Authentic Foods or Mochiko brand from Koda Farms: 150 grams per cup
- Thai glutinous (aka sweet) white rice flour (Erawan): 120 grams per cup
- Brown rice flour from Bob's Red Mill or superfine brown rice flour from Authentic Foods: 135 grams per cup

Sorghum Flour

- Sorghum flour from Bob's Red Mill or superfine sorghum flour from Authentic Foods: 135 to 140 grams per cup

Teff Flour

- Teff flour from Bob's Red Mill: 135 grams per cup

DIY Nut Flour

Homemade nut flour is not only fresher, it is more flavorful and often (but not always) less expensive than purchased flour. Nut flours can be made with a food processor, a blender, a hand-crank nut grinder, or even a coffee grinder. The texture—how coarse or fine your flour will be—depends on the equipment you use. With some equipment, you can even make the exact amount of flour you need when you need it with no guessing or waste: just weigh out the nuts, then make them into flour. Homemade nut flour varies more widely in texture and weight per cup than purchased nut flours, so if you make flour to keep on hand, measure it by weight rather than volume when following a recipe.

Nut flour should be dry rather than oily or pasty to the touch. Food processors, blenders, and coffee grinders—all of which keep the nuts in continuous contact with a blade—require vigilance to avoid an oily meal or an outright nut butter. With a little care, hard nuts like almonds and hazelnuts can be made successfully in this type of equipment (especially in the coffee grinder or blender rather than a food processor). Soft, oily nuts like walnuts and pecans are harder to control; they become oily in the blink of an eye—a little bit of flour or sugar from the recipe added to the processor can help slightly, but you must still take care.

The best equipment for pulverizing any nuts, but especially for soft nuts, involves a shredding or grating disk or drum instead of a blade—a small hand-cranked, table-mounted nut grinder with barrel shaped grater is excellent, as is the fine grating disk of a food processor—the nuts pass through just once and the resulting flour is always fine, dry, and fluffy.

Before you start, nuts should be at room temperature rather than hot from the oven, and dry (not moist from defrosting). Equipment should also be dry and at room temperature rather than warm or hot from the dishwasher.

Here are some equipment choices with notes on which nuts work best with what.

Food Processor with Steel Blade

Makes a relatively coarse flour in comparison with purchased flour. Okay for almonds and hazelnuts if you have nothing else, or want or don't mind a coarse texture. Soft nuts get oily.

Fill the bowl only ¼ to ⅓ full. Stop and scrape around the corners of the bowl from time to time, until the nuts are relatively evenly pulverized but not oily.

(continues)

Food Processor with Fine Shredder Blade

Makes fine, fluffy flour with hard or soft nuts, but a high percentage of large pieces bypass the disk and must be strained and reserved for another use or else reprocessed. It's impossible to process the exact amount needed.

Attach the fine shredder blade to the processor. Use the feed tube. Turn the processor on before adding nuts. Continue adding nuts to the tube, letting them pass through the shredder without pushing. Shake the flour through a coarse sieve to remove the larger pieces. Reserve the pieces for another use or reprocess them with the steel blade and stir the resulting meal into the nut flour.

Blender

Makes excellent medium-fine almond or hazelnut flour (better than with a food processor with steel blade). Soft nuts get oily: be content with coarser flour.

Pulse about 1 cup of nuts at time, scraping the bowl from time to time.

Coffee Grinder

Makes excellent fine almond or hazelnut flour. Soft nuts are hard to control and get oily fast.

Fill the grinder only about half full. Pulse until fine but not oily.

Nut Grinder (table-mounted with a barrel-shaped grater and a hand crank)

Makes superb flour that is finer and fluffier than any purchased flour. It's easy to make the exact amount of flour (by weight) needed for a recipe. If you make extra flour to have on hand, always measure it by weight rather than volume for recipes.

ACKNOWLEDGMENTS

I thank Maya Klein for her inspired work and dedication to our first "official" book together—after years of collaboration and help behind the scenes. Thanks to our editor, Judy Pray, at Artisan, for seeing the big picture and encouraging us to organize the book by flour rather than type of recipe. Thanks to the late, great Peter Workman, Ann Bramson, and the rest of the Artisan team: Sibylle Kazeroid, Bridget Monroe Itkin, Allison McGeehon, Michelle Ishay-Cohen, Lia Ronnen, and Nancy Murray, as well as Sarah Weaver and Laura Klynstra.

Our families tasted endless samples and gave brave feedback: thanks to Maya's husband, Steve Klein, and her sons, Nate and Wade; my daughter, Lucy Medrich; my brother and sister-in-law, Albert and Tami Abrams; and my mother, Bea Abrams.

Leigh Beisch's photos are gorgeous; Sandra Cook's styling is divine. Thanks to Penny Flood for her calm reign over the studio kitchen and that exquisitely styled fudge cake! Stylist and creative director Sara Slavin was also captain of our studio team for two intensely productive weeks. I salute her yet again. Thanks always to my agent, Jane Dystel, for her support and encouragement.

INDEX BY HERO FLOUR

Note: Page numbers in *italics* refer to illustrations.

buckwheat flour, 166–95
- Better than Buckwheat Blini, 185–86, *187*
- Buckwheat Cake with Rose Apples, 172, *173*, 174
- Buckwheat Coffee Baby Cakes with Toffee Sauce, 181, *182*, 183
- Buckwheat Gingerbread, 175
- Buckwheat Linzer Cookies, 190, *191*
- Buckwheat Sablés, 189
- Buckwheat Sour Cream Soufflés with Honey, *192*, 193
- Buckwheat Sponge Cake, *170*, 171
- Buckwheat Walnut or Hazelnut Tuiles, 188
- Dark and Spicy Pumpkin Loaf, 178, *179*, 180
- Date-Nut Cake with Cherries and Buckwheat, 177
- flavor affinities, 169
- Panforte Nero, 184
- Walnut and Buckwheat Crackers, 194
- weight and volume, 353
- where to buy and how to store, 169

chestnut flour, 196–227
- Chestnut and Pine Nut Shortbread, 217
- Chestnut and Walnut Meringues, 218–19
- Chestnut Bûche de Noël, 202–4, *205*
- Chestnut Jam Tart, 212, *213*
- Chestnut Meringues Glacées, *220*, 221–22
- Chestnut Praline Gelato, 227
- Chestnut Pudding, 226
- Chestnut Sponge Cake with Pear Butter and Crème Fraîche, 200–201
- Chocolate Chestnut Soufflé Cake, 206–7, *207*
- flavor affinities, 199
- Quince and Orange-Filled Chestnut Cookies, 223–24, *225*
- Ricotta Cheesecake with Chestnut Crust, *214*, 215–16
- Walnut and Honey Tart with Chestnut Crust, *208*, 209–11
- weight and volume, 353
- where to buy and how to store, 199

coconut and coconut flour, 294–331
- Carrot Coconut Almond Torte, 307–8
- Chocolate Coconut Tart, 318
- Chunky Double-Chocolate Coconut Meringues, 326, *327*
- Coconut and Mango Tart with Coconut Pastry Cream, 315
- Coconut Chiffon Cake, 303
- Coconut Chocolate Pecan Torte, 310, *311*
- Coconut Key Lime Tart, *316*, 317
- Coconut Marjolaine, 312, *313*, 314
- Coconut Meringues, 328
- Crispy Coconut Wafers, 324, *325*
- Crunchy Coconut Cookies, 321
- flavor affinities, 297
- Walnut Alfajores, *322*, 323
- weight and volume, 354
- where to buy and how to store, 297

corn flour and cornmeal, 136–65
- Blueberry Corn Flour Cobbler, 158, *159*
- Corn Flour and Cranberry Scones, *150*, 151
- Corn Flour Biscuits, 147
- Corn Flour Chiffon Cake, 140–41
- Corn Flour Tea Cake with Currants and Pistachios, 148, *149*
- Crunchy Corn Fritters, 152, *153*
- flavor affinities, 139
- Golden Corn Cake, 142–43
- Lemon Tart, 154–55
- The New Chocolate Cream Pie, *156*, 157
- New Classic Boston Cream Pie, *144*, 145–46
- Seed Crackers, 163
- Souffléed Corn Flour and Yogurt Puddings with Cajeta, 164, *165*
- stone-ground cornmeal, 37
- Sweet or Savory Corn Sticks, *160*, 161–62
- weight and volume, 354
- where to buy and how to store, 139

nut flours, 294–331
- Almond Butter Cake with Spiced Apricots, 300, *301*, 302
- Almond Drop Biscuits, 320
- Carrot Coconut Almond Torte, 307–8
- DIY nut flour, 355–56
- flavor affinities, 297
- Hazelnut Layer Cake with Dark Chocolate Frosting and Blackberry Preserves, 304
- Nutty Sponge Cake, *298*, 299
- Peach-Almond Cobbler, 319
- Pistachio Butter Cake with Sweet Pickled Peaches, 305–6
- weight and volume, 354
- where to buy and how to store, 297

oat flour, 82–135
- Apple Crumble, 116

Apricot Walnut Rugelach, *120*, 121–23
Blackberry Galette, 113
Butter Biscuits, *108*, 109
Buttery Apple Cobbler, 117
Caramel Upside-Down Cake, 96, *97*, 98
Carrot Spice Cake with Cream Cheese Frosting, *100*, 101–2
Chocolate Chip Cookies, 126–27
Chocolate Layer Cake, 94–95
Chocolate Sheet Cake, 106
Cinnamon Crumb Cake, 99
Classic Ginger Cookies, 125
Cutout Cookies, *130*, 131–32
Double Oatmeal Cookies, 124
flavor affinities, 85
Maya's Chocolate Fudge Cake with Milk Chocolate Frosting, 103–4, *105*
New Classic Blondies, 118
Nutty Thumbprint Cookies, 133–34, *135*
Oat and Almond Tuiles, 119
Oat Flour Fruit Basket Cake, 88, *89*
Oat Flour Sponge Cake, 86–87, *87*
Oat Sablés, 128–29
Peach Crumble, *114*, 115
Poppy Seed Pound Cake, *92*, 93
Simple Scones, 107
Ultimate Butter Cake, 90–91
weight and volume, 354
where to buy and how to store, 85
Yogurt Tart, 110, *111*, 112

rice flour, 50–81
Almond and Brown Rice Brownies, 71
Almond Tuiles, 67, *68*, 69
Beignets, *64*, 65–66
Brandied Cherry Clafoutis, 72, *73*
brown, 52
Brown Rice Sponge Cake with Three Milks, 57, *58*, 59
Dark Chocolate Soufflés, *74*, 75–76
flavor affinities, 53
glutinous, 52
Hazelnut Crunch Brownies, 330
Lemon Cream Roulade with Strawberry-Mint Salad, 60, *61*, 62–63
Peanut Crunch Brownies, 330, *331*
Sicilian Chocolate Gelato, 81
Silky Butterscotch Pudding, 77
Silky Chocolate Pudding, 80
Silky Saffron Rice Pudding, 78, *79*
as thickener, 77
Ultra-Bittersweet Brownies, 70
weight and volume, 354
where to buy and how to store, 53
white, 52
White Rice Chiffon Cake, *54*, 55–56

sorghum flour, 260–93
Banana Muffins, *268*, 269
Fig and Anise Scones, 270, *271*
flavor affinities, 263
Ginger-Peach Squares, 285, *286*, 287
Nutella Sandwich Cookies, 288, *289*
Nutmeg Shortbread, 283
Salted Peanut Shorties, 284
Sorghum Cinnamon Sticks, 290–91
Sorghum Ice Cream with Peanut Brittle, 292, *293*
Sorghum Layer Cake with Walnut Praline Buttercream, *264*, 265–67
Sorghum Pecan Tart, 277, *278*, 279
Strawberry-Blueberry Tart, 275–76
Strawberry Tartlets, *272*, 273–74
Toasty Pecan Biscotti with Hibiscus Pears, 280, *281*, 282
weight and volume, 354
where to buy and how to store, 263

teff flour, 228–59
Bittersweet Teff Brownies, *256*, 257
Black Cherry Chocolate Linzer Torte, 245–46
Chocolate Raspberry Celebration Cake, 240–42, *243*
Chocolate Sablés, 250–51
Cocoa Crepes filled with Chocolate and Walnuts, *252*, 253–54
Date-Nut Cake with Apricots and Teff, *176*, 244
flavor affinities, 231
German Chocolate Cake, 237
Hungarian Crepe Cake, 255
Mocha Cream Tart with Chocolate Crust, 247–48, *249*
The New Chocolate Génoise, 232–33
Queen of the Nile, *238*, 239
Tangy Aromatic Crackers, *195*, 259
Teff Blini, 258
Triple Chocolate Layer Cake, 234–36
weight and volume, 353
where to buy and how to store, 231

GENERAL INDEX

Note: Page numbers in *italics* refer to illustrations.

Alfajores, Walnut, *322*, 323
almonds, *see* nuts
anise: Fig and Anise Scones, 270, *271*
apples:
  Apple Crumble, 116
  Buckwheat Cake with Rose Apples, 172, *173*, 174
  Buttery Apple Cobbler, 117
  Caramel Apple Upside-Down Cake, 96, *97*, 98
apricots:
  Almond Butter Cake with Spiced Apricots, 300, *301*, 302
  Apricot Crumble, 115
  Apricot Walnut Rugelach, *120*, 121–23
  Date-Nut Cake with Apricots and Teff, 244
  Date-Nut Cake with Cherries and Buckwheat, *176*, 177

baking powder, 32
baking soda, 32
bananas:
  Banana Muffins, *268*, 269
  Nutella-Filled Teff Crepes (with or without bananas), 254
Beignets, *64*, 65–66
Biscotti, Toasty Pecan, 280
biscuits:
  Almond Drop Biscuits, 320
  Blueberry Corn Flour Cobbler, 158, *159*
  Butter Biscuits, *108*, 109
  Buttery Apple Cobbler, 117
  Corn Flour Biscuits, 147
  Peach-Almond Cobbler, 319
Blackberry Corn Flour Cobbler, 158
Blackberry Crumble, 115
Blackberry Galette, 113
blini:
  Better than Buckwheat Blini, 185–86, *187*
  Teff Blini, 258
Blondies, New Classic, 118
Blueberry, Blackberry, or Huckleberry Crumble, 115
Blueberry Corn Flour Cobbler, 158, *159*
Blueberry Walnut Rugelach, 123
Boston Cream Pie, New Classic, *144*, 145–46
Bourbon Glaze, 91
brandy:
  Brandied Cherry Clafoutis, 72, *73*
  Prunes Poached in Coffee and Brandy, 336
brownies:
  Almond and Brown Rice Brownies, 71
  Bittersweet Teff Brownies, *256*, 257
  Hazelnut Crunch Brownies, 330
  Peanut Crunch Brownies, 330, *331*
  Ultra-Bittersweet Brownies, 70
Bûche de Noël, Chestnut, 202–4, *205*
butter, 32
  Almond Butter Cake with Spiced Apricots, 300, *301*, 302
  Clarified Butter (Ghee), 38, 335
  New Classic Buttercream, 348
  Pistachio Butter Cake with Sweet Pickled Peaches, 305–6
  Poppy Seed Pound Cake, *92*, 93
  Ultimate Butter Cake, 90–91

cacao nibs, 33
  Cacao Nib Rugelach, 123
  Nibby Nut and Raisin Cookies, 127
  Nibby Oat Sablés, 129
  Nibby Walnut Shorties, 284
cajeta, 38
  Souffléed Corn Flour and Yogurt Puddings with Cajeta, 164, *165*
cakes:
  Almond Butter Cake with Spiced Apricots, 300, *301*, 302
  Brown Rice Sponge Cake with Three Milks, 57, *58*, 59
  Buckwheat Cake with Rose Apples, 172, *173*, 174
  Buckwheat Coffee Baby Cakes with Toffee Sauce, 181, *182*, 183
  Buckwheat Sponge Cake, *170*, 171
  Caramel Apple Upside-Down Cake, 96, *97*, 98
  Cardamom and Saffron Rice Chiffon Cake, 56
  Carrot Spice Cake with Cream Cheese Frosting, *100*, 101–2
  Chestnut Bûche de Noël, 202–4, *205*
  Chestnut Sponge Cake with Pear Butter and Crème Fraîche, 200–201
  Chocolate Chestnut Soufflé Cake, 206–7, *207*
  Chocolate Layer Cake, 94–95
  Chocolate Raspberry Celebration Cake, 240–42, *243*
  Chocolate Sheet Cake, 106
  Cinnamon Crumb Cake, 99
  Coconut Chiffon Cake, 303
  Coconut Marjolaine, 312, *313*, 314
  Corn Flour Chiffon Cake, 140–41
  Corn Flour Tea Cake with Currants and Pistachios, 148, *149*
  Dark and Spicy Pumpkin Loaf, 178, *179*, 180

Date-Nut Cake with Apricots and Teff, *176*, 244
Date-Nut Cake with Cherries and Buckwheat, 177
Figgy Upside-Down Cake, 98
German Chocolate Cake, 237
Golden Corn Cake, 142–43
Hazelnut Layer Cake with Dark Chocolate Frosting and Blackberry Preserves, 304
Holiday Pound Cake with Bourbon Glaze, 91
Hungarian Crepe Cake, 255
Lemon and Corn Flour Chiffon Cake, 141
Lemon Cream Roulade with Strawberry-Mint Salad, 60, *61*, 62–63
Lemon Rice Chiffon Cake, 56
Maya's Chocolate Fudge Cake with Milk Chocolate Frosting, 103–4, *105*
New Classic Boston Cream Pie, *144*, 145–46
Nondairy Fudge Cake, 104
Nutty Sponge Cake, *298*, 299
Oat Flour Fruit Basket Cake, 88, *89*
Oat Flour Sponge Cake, 86–87, *87*
Orange Butter Cake, 91
Orange Rice Chiffon Cake, 56
Pistachio Butter Cake with Sweet Pickled Peaches, 305–6
Poppy Seed Pound Cake, *92*, 93
Queen of the Nile, *238*, 239
Red Velvet–Style Cake, 104
removing skins from, 235
Sorghum Layer Cake with Walnut Praline Buttercream, *264*, 265–67
Sweet Potato Cake, 101
Triple Chocolate Layer Cake, 234–36
Ultimate Butter Cake, 90–91
White Rice Chiffon Cake, *54*, 55–56
White Rice Sponge Cake (aka Génoise), 63
Caramel Apple Upside-Down Cake, 96, *97*, 98
Cardamom and Saffron Rice Chiffon Cake, 56
Cardamom Brittle, 338–39
Carrot Coconut Almond Torte, 307–8
Carrot Spice Cake with Cream Cheese Frosting, *100*, 101–2
Caviar, Blini with, 186, *187*
Chai Sablés, 129
cheesecake: Ricotta Cheesecake with Chestnut Crust, *214*, 215–16
cherries:
Black Cherry Chocolate Linzer Torte, 245–46
Brandied Cherry Clafoutis, 72, *73*
Cherry Crumble, 115
Date-Nut Cake with Cherries and Buckwheat, 177
Sautéed Cherries, 143
Chestnut Bûche de Noël, 202–4, *205*
Chestnut Meringue Mushrooms, 219
Chestnut Pudding, 226
chocolate, 32–33, 37
Almond and Brown Rice Brownies, 71
Bittersweet Teff Brownies, *256*, 257
Black Cherry Chocolate Linzer Torte, 245–46
Brown Butter Chocolate Génoise, 233
Chestnut Bûche de Noël, 202–4, *205*
Chestnut Meringue Mushrooms, 219
Chocolate Chestnut Soufflé Cake, 206–7, *207*
Chocolate Chip Cookies, 126–27
Chocolate Coconut Tart, 318
Chocolate Fudge Cupcakes, 104
Chocolate-Glazed Beignets, 66
Chocolate-Hazelnut Rugelach, 123
Chocolate Hazelnut Tart, 318
Chocolate Layer Cake, 94–95
Chocolate-Mint Sandwich Cookies, 251
Chocolate Raspberry Celebration Cake, 240–42, *243*
Chocolate Sablés, 250–51
Chocolate Sheet Cake, 106
Chunky Double-Chocolate Coconut Meringues, 326, *327*
Cocoa Crepes Filled with Chocolate and Walnuts, *252*, 253–54
Coconut Chocolate Pecan Torte, 310, *311*
Crunchy Coconut Cookies with Bittersweet Chocolate, 321
Dark Chocolate Frosting, 95
Dark Chocolate Soufflés, *74*, 75–76
German Chocolate Cake, 237
Hazelnut Crunch Brownies, 330
Hazelnut Layer Cake with Dark Chocolate Frosting and Blackberry Preserves, 304
how to melt, 34–35, *36*
Maya's Chocolate Fudge Cake with Milk Chocolate Frosting, 103–4, *105*
Mocha Cream Tart with Chocolate Crust, 247–48, *249*
Mocha Mousse Frosting, 346
The New Chocolate Cream Pie, *156*, 157
The New Chocolate Génoise, 232–33
New Classic Blondies, 118
Nondairy Fudge Cake, 104
Nutella Sandwich Cookies, 288, *289*
Peanut Crunch Brownies, 330, *331*
Queen of the Nile, *238*, 239
Sarah Bernhardt Chocolate Glaze, 347
Sicilian Chocolate Gelato, 81
Silky Chocolate Pudding, 80
Souffléed Corn Flower Puddings Laced with Milk Chocolate, 164, *165*
Spicy Chocolate Sablés, 251
Triple Chocolate Layer Cake, 234–36
Ultra-Bittersweet Brownies, 70
Whipped Ganache Filling, 236
cinnamon:
Beignets with Cinnamon Sugar, 66
Cinnamon Crumb Cake, 99
Cinnamon-Orange Brittle, 339
Cinnamon Sugar, 290
Sorghum Cinnamon Sticks, 290–91
Clafoutis, Brandied Cherry, 72, *73*

cocoa powder, 32–33
Cocoa Crepes Filled with Chocolate and Walnuts, *252*, 253–54
Cocoa Fudge Glaze, 146
Cocoa Teff Brownies, 257
coconut, 37
Carrot Coconut Almond Torte, 307–8
Chocolate Coconut Tart, 318
Chunky Double-Chocolate Coconut Meringues, 326, *327*
Coconut and Mango Tart with Coconut Pastry Cream, 315
Coconut Chiffon Cake, 303
Coconut Chocolate Pecan Torte, 310, *311*
Coconut Key Lime Tart, *316*, 317
Coconut Marjolaine, 312, *313*, 314
Coconut Meringues, 328
Coconut Pastry Cream, 342
Crispy Coconut Wafers, 324, *325*
Crunchy Coconut Cookies, 321
German Chocolate Cake, 237
Oat and Coconut Tuiles, 119
toasting, 309
coffee, 37
Buckwheat Coffee Baby Cakes with Toffee Sauce, 181, *182*, 183
Coffee Walnut Shorties, 284
Coffee Whipped Cream, 345
Mocha Cream Tart with Chocolate Crust, 247–48, *249*
Mocha Mousse Frosting, 346
Prunes Poached in Coffee and Brandy, 336
Sorghum Layer Cake with Coffee-Walnut Praline Buttercream, 267
cookies:
Almond Tuiles, 67, *68*, 69
Apricot Walnut Rugelach, *120*, 121–23
Brown Sugar Pecan Cookies, 329
Buckwheat Linzer Cookies, 190, *191*
Buckwheat Sablés, 189
Buckwheat Walnut or Hazelnut Tuiles, 188
Chestnut and Pine Nut Shortbread, 217
Chestnut and Walnut Meringues, 218–19
Chestnut Meringues Glacées, *220*, 221–22
Chocolate Chip Cookies, 126–27
Chocolate-Mint Sandwich Cookies, 251
Chocolate Sablés, 250–51
Chunky Double-Chocolate Coconut Meringues, 326, *327*
Classic Ginger Cookies, 125
Coconut Meringues, 328
Coffee Walnut Shorties, 284
Crispy Coconut Wafers, 324, *325*
Crunchy Coconut Cookies, 321
Cutout Cookies, *130*, 131–32
Double Oatmeal Cookies, 124
Ginger-Peach Squares, 285, *286*, 287
Molasses Spice Cookies, 125
New Classic Blondies, 118
Nibby Nut and Raisin Cookies, 127
Nibby Walnut Shorties, 284
Nutella Sandwich Cookies, 288, *289*
Nutmeg Shortbread, 283
Nutty Peanut Cookies, 329
Nutty Thumbprint Cookies, 133–34, *135*
Oat Sablés, 128–29
Pecan Spice Cookies, 127
Quince and Orange–Filled Chestnut Cookies, 223–24, *225*
Salted Peanut Shorties, 284
Sorghum Cinnamon Sticks, 290–91
Spicy Basil Sablés, 129
Sweet or Savory Corn Sticks, *160*, 161–62
Toasty Pecan Biscotti with Hibiscus Pears, 280, *281*, 282
Walnut Alfajores, *322*, 323
Walnut or Pecan Sablés, 134
Corn Fritters, Crunchy, 152, *153*
Corn Sticks, Sweet or Savory, *160*, 161–62
crackers:
Seed Crackers, 163
Tangy Aromatic Crackers, *195*, 259
Walnut and Buckwheat Crackers, 194
cranberries: Corn Flour and Cranberry Scones, *150*, 151
cream, 37
Almond Drop Biscuits, 320
Buckwheat Cake with Rose Apples, 172, *173*, 174
Buckwheat Coffee Baby Cakes with Toffee Sauce, 181, *182*, 183
Butter Biscuits, *108*, 109
Chestnut Praline Gelato, 227
Chestnut Pudding, 226
Chocolate Coconut Tart, 318
Coconut Key Lime Tart, *316*, 317
Coconut Marjolaine, 312, *313*, 314
Corn Flour and Cranberry Scones, *150*, 151
Corn Flour Biscuits, 147
Fig and Anise Scones, 270, *271*
Lemon Cream Roulade with Strawberry-Mint Salad, 60, *61*, 62–63
Mocha Cream Tart with Chocolate Crust, 247–48, *249*
Mocha Mousse Frosting, 346
Oat Flour Fruit Basket Cake, 88, *89*
Silky Butterscotch Pudding, 77
Silky Saffron Rice Pudding, 78, *79*
Simple Scones, 107
Sorghum Ice Cream with Peanut Brittle, 292, *293*
Toffee Sauce, *182*, 183
Walnut and Honey Tart with Chestnut Crust, *208*, 209–11
Whipped Cream, 343
Whipped Ganache Filling, 236
cream cheese:
Apricot Walnut Rugelach, *120*, 121–23
Cream Cheese Frosting, *100*, 102

crème fraîche:
Chestnut Sponge Cake with Pear Butter and Crème Fraîche, 200–201
Chocolate Raspberry Celebration Cake, 240–42, *243*
Whipped Crème Fraîche, 345
crepes, 21, 25, 26
Basic Crepes, 24
Cocoa Crepes Filled with Chocolate and Walnuts, *252*, 253–54
Hungarian Crepe Cake, 255
Nutella-Filled Teff Crepes, 254
cupcakes:
Chocolate Fudge Cupcakes, 104
Ultimate Butter Cupcakes, 91
curd: Lemon or Lime Curd, 341
currants:
Cacao Nib Rugelach, 123
Corn Flour Tea Cake with Currants and Pistachios, 148, *149*
Cutout Cookies, *130*, 131–32

dates:
Buckwheat Coffee Baby Cakes with Toffee Sauce, 181, *182*, 183
Date-Nut Cake with Apricots and Teff, 244
Date-Nut Cake with Cherries and Buckwheat, *176*, 177
Date-Nut Scones, 107
Double Oatmeal Cookies, 124
dulce de leche, 38
Walnut Alfajores, *322*, 323

eggs:
Buckwheat Sponge Cake, *170*, 171
Chestnut Bûche de Noël, 202–4, *205*
Coconut Chocolate Pecan Torte, 310, *311*
Coconut Marjolaine, 312, *313*, 314
Corn Flour Chiffon Cake, 140–41
Golden Corn Cake, 142–43
Nutty Sponge Cake, *298*, 299
White Rice Chiffon Cake, *54*, 55–56
egg wash, 209, 210
equipment, 40–48

figs:
Chestnut Sponge Cake with Figs, 201
Fig and Anise Scones, 270, *271*
Figgy Upside-Down Cake, 98
Panforte Nero, 184
Sweet Pickled Figs, 306
flours:
buying and storing, 27
fineness of, 27–28
measuring with cups, 29, *31*
weight and volume, 28–29, 353–54
see also Index by Hero Flour
Fritters, Crunchy Corn, 152, *153*

frostings:
buttercream tips, 349
Coconut Marjolaine, 312, *313*, 314
Cream Cheese Frosting, *100*, 102
Dark Chocolate Frosting, 95
Milk Chocolate Frosting, 104
Mocha Mousse Frosting, 346
New Classic Buttercream, 348
fruit, dried, 38
fruit preserves:
Apricot Walnut Rugelach, *120*, 121–23
Black Cherry Chocolate Linzer Torte, 245–46
Buckwheat Linzer Cookies, 190, *191*
Chestnut Jam Tart, 212, *213*
Hazelnut Layer Cake with Dark Chocolate Frosting and Blackberry Preserves, 304
Hungarian Crepe Cake, 255
Oat Flour Fruit Basket Cake, 88, *89*

Galette, Blackberry, 113
Ganache Filling, Whipped, 236
génoise:
Brown Butter Chocolate Génoise, 233
The New Chocolate Génoise, 232–33
removing skins from, 235
White Rice Sponge Cake Génoise, 63
German Chocolate Cake, 237
ghee (clarified butter), 38, 335
ginger:
Buckwheat Gingerbread, 175
Classic Ginger Cookies, 125
Ginger-Peach Squares, 285, *286*, 287
glazes:
Bourbon Glaze, 91
Chocolate Glaze, 66
Cocoa Fudge Glaze, 146
Red Currant Jelly Glaze, 276
Sarah Bernhardt Chocolate Glaze, 347
Golden Corn Cake, 142–43
Grapefruit and Basil Sablés, 129

Hazelnut Layer Cake with Dark Chocolate Frosting and Blackberry Preserves, 304
Hibiscus Pears, *281*, 282
Holiday Pound Cake with Bourbon Glaze, 91
honey:
Better than Buckwheat Blini, 185–86, *187*
Buckwheat Sour Cream Soufflés with Honey, *192*, 193
Honey Whipped Cream, 345
Panforte Nero, 184
Spiced Apricots, 302
Teff Blini, 258
Walnut and Honey Tart with Chestnut Crust, *208*, 209–11
Hungarian Crepe Cake, 255
ice cream:
Chestnut Meringues Glacées, *220*, 221–22
Chestnut Praline Gelato, 227

ice cream *(cont.)*
Sicilian Chocolate Gelato, 81
Sorghum Ice Cream with Peanut Brittle, 292, *293*

jams, *see* fruit preserves

lemon:
Individual Lemon Roulades, 63
Lemon and Corn Flour Chiffon Cake, 141
Lemon Cream Roulade with Strawberry-Mint Salad, 60, *61*, 62–63
Lemon Curd, 341
Lemon Custard Pastry Cream, 342
Lemon Rice Chiffon Cake, 56
Lemon Tart, 154–55
lime:
Coconut Key Lime Tart, *316*, 317
Lime and Mint Sablés, 129
Lime Curd, 341
Linzer Cookies, Buckwheat, 190, *191*
Linzer Torte, Black Cherry Chocolate, 245–46

mango: Coconut and Mango Tart with Coconut Pastry Cream, 315
Maya's Chocolate Fudge Cake with Milk Chocolate Frosting, 103–4, *105*
measuring cups and spoons, 29, 42, 44
meringues:
Chestnut and Walnut Meringues, 218–19
Chestnut Meringue Mushrooms, 219
Chestnut Meringues Glacées, *220*, 221–22
Chunky Double-Chocolate Coconut Meringues, 326, *327*
Coconut Marjolaine, 312, *313*, 314
Coconut Meringues, 328
milk:
Brown Rice Sponge Cake with Three Milks, 57, *58*, 59
Coconut Key Lime Tart, *316*, 317
Silky Chocolate Pudding, 80
mint:
Chocolate-Mint Sandwich Cookies, 251
Lime and Mint Sablés, 129
Strawberry-Mint Salad, 60, *61*, 62–63
Mocha Cream Tart with Chocolate Crust, 247–48, *249*
Mocha Mousse Frosting, 346
molasses:
Buckwheat Gingerbread, 175
Classic Ginger Cookies, 125
Molasses Spice Cookies, 125
muffins:
Banana Muffins, *268*, 269
Dark and Spicy Pumpkin Muffins, 180
Mushrooms, Chestnut Meringue, 219

Nectarine Crumble, 115
Nectarine Upside-Down Cake, 98
Nondairy Fudge Cake, 104
Nutella-Filled Teff Crepes (with or without bananas), 254
Nutella Sandwich Cookies, 288, *289*
Nutmeg Shortbread, 283
Nutmeg Sugar, 283
nuts, 38
Almond and Brown Rice Brownies, 71
Almond Tuiles, 67, *68*, 69
Apple Crumble, 116
Apricot Walnut Rugelach, *120*, 121–23
Banana Muffins, *268*, 269
Bittersweet Teff Brownies, *256*, 257
Black Cherry Chocolate Linzer Torte, 245–46
Blueberry Walnut Rugelach, 123
Brown Sugar Pecan Cookies, 329
Buckwheat Walnut or Hazelnut Tuiles, 188
Cardamom Brittle, 338–39
Carrot Spice Cake with Cream Cheese Frosting, *100*, 101–2
Chestnut and Pine Nut Shortbread, 217
Chestnut and Walnut Meringues, 218–19
Chestnut Jam Tart, 212, *213*
Chestnut Sponge Cake with Praline Whipped Cream, 201
Chocolate Chestnut Soufflé Cake, 206–7, *207*
Chocolate Chip Cookies, 126–27
Chocolate-Hazelnut Rugelach, 123
Cinnamon Crumb Cake, 99
Cinnamon-Orange Brittle, 339
Cocoa Crepes Filled with Chocolate and Walnuts, *252*, 253–54
Coconut Chocolate Pecan Torte, 310, *311*
Coffee Walnut Shorties, 284
Corn Flour Tea Cake with Currants and Pistachios, 148, *149*
Date-Nut Cake with Apricots and Teff, 244
Date-Nut Cake with Cherries and Buckwheat, *176*, 177
Date-Nut Scones, 107
DIY Nut Flour, 355–56
Double Oatmeal Cookies, 124
German Chocolate Cake, 237
Hazelnut Crunch Brownies, 330
Hungarian Crepe Cake, 255
New Classic Blondies, 118
Nibby Nut and Raisin Cookies, 127
Nibby Walnut Shorties, 284
Nutella Sandwich Cookies, 288, *289*
Nutty Oat Sablés, 129
Nutty Peanut Cookies, 329
Nutty Thumbprint Cookies, 133–34, *135*
Oat and Almond Tuiles, 119
Panforte Nero, 184
Peach Crumble, *114*, 115
Peanut Brittle, 292, *293*
Peanut Crunch Brownies, 330, *331*
Pecan Spice Cookies, 127
Ricotta Cheesecake with Chestnut Crust, *214*, 215–16

Salted Peanut Shorties, 284
Sorghum Layer Cake with Coffee-Walnut Praline Buttercream, 267
Sorghum Layer Cake with Praline Whipped Cream, 267
Sorghum Layer Cake with Walnut Praline Buttercream, *264*, 265–67
Sorghum Pecan Tart, *277*, *278*, 279
toasting, 309
Toasty Pecan Biscotti with Hibiscus Pears, 280, *281*, 282
Walnut Alfajores, *322*, 323
Walnut and Buckwheat Crackers, 194, *195*
Walnut and Honey Tart with Chestnut Crust, *208*, 209–11
Walnut or Pecan Praline Brittle, 337
Walnut or Pecan Sablés, 134
Nutty Sponge Cake, *298*, 299

oats: Double Oatmeal Cookies, 124
orange juice or zest:
Cinnamon-Orange Brittle, 339
Orange Butter Cake, 91
Orange Rice Chiffon Cake, 56
Orange Sablés with Ancho Chile, 129
Orange-Saffron Sablés, 129
Quince and Orange–Filled Chestnut Cookies, 223–24, *225*
oven rack position, 29

pancakes, 21–22
Basic Pancakes, 23
Better than Buckwheat Blini, 185–86, *187*
Teff Blini, 258
Panforte Nero, 184
pans, 41, 42, 47; lining, 31
Pastry Cream, New Vanilla, 342
peaches:
Ginger-Peach Squares, 285, *286*, 287
Peach-Almond Cobbler, 319
Peach Cobbler, 158
Peach Crumble, *114*, 115
Pistachio Butter Cake with Sweet Pickled Peaches, 305–6
pears:
Hibiscus Pears, *281*, 282
Pear and Rhubarb Cobbler, 117
Pear Clafoutis, 72
Pear Crumble, 116
Pear Upside-Down Cake, 98
Pistachio Butter Cake with Sweet Pickled Peaches, 305–6
plums:
Plum Crumble, 115
Prune Plum Cobbler, 158
Poppy Seed Pound Cake, *92*, 93
potato starch, 38
praline:
Chestnut Praline Gelato, 227
Chestnut Sponge Cake with Praline Whipped Cream, 201
Praline Whipped Cream, 345
Sorghum Layer Cake with Coffee-Walnut Praline Buttercream, 267
Sorghum Layer Cake with Praline Whipped Cream, 267
Sorghum Layer Cake with Walnut Praline Buttercream, *264*, 265–67
Walnut or Pecan Praline Brittle, 337
Prunes Poached in Coffee and Brandy, 336
puddings:
Chestnut Pudding, 226
The New Chocolate Cream Pie, *156*, 157
Silky Butterscotch Pudding, 77
Silky Chocolate Pudding, 80
Silky Saffron Rice Pudding, 78, *79*
Silky Vanilla Pudding, 78
Souffléed Corn Flour and Yogurt Puddings with Cajeta, 164, *165*
thickeners for, 77
pumpkin: Dark and Spicy Pumpkin Loaf, 178, *179*, 180

Queen of the Nile, *238*, 239
Quince and Orange–Filled Chestnut Cookies, 223–24, *225*

raisins:
Double Oatmeal Cookies, 124
Nibby Nut and Raisin Cookies, 127
Red Velvet–Style Cake, 104
rhubarb: Pear and Rhubarb Cobbler, 117
Ricotta Cheesecake with Chestnut Crust, *214*, 215–16
Rose Apples, 174
Rose Whipped Cream, 345
rugelach: Apricot Walnut Rugelach, *120*, 121–23

saffron:
Cardamom and Saffron Rice Chiffon Cake, 56
Orange-Saffron Sablés, 129
Silky Saffron Rice Pudding, 78, *79*
Salad, Strawberry-Mint, 60, *61*, 62–63
Salmon, Smoked, Blini with, 186, *187*
salt, 39
Salted Peanut Shorties, 284
Sarah Bernhardt Chocolate Glaze, 347
scones:
Almond Drop Scones, 320
Corn and Blueberry Scones, 151
Corn Flour and Cranberry Scones, *150*, 151
Date-Nut Scones, 107
Fig and Anise Scones, 270, *271*
Simple Scones, 107
Seed Crackers, 163
Sicilian Chocolate Gelato, 81
Souffléed Corn Flour and Yogurt Puddings with Cajeta, 164, *165*

soufflés:
- Buckwheat Sour Cream Soufflés with Honey, *192*, 193
- Chocolate Chestnut Soufflé Cake, 206–7, *207*
- Dark Chocolate Soufflés, *74*, 75–76

spices, 39
- Buckwheat Gingerbread, 175
- Carrot Spice Cake with Cream Cheese Frosting, *100*, 101–2
- Classic Ginger Cookies, 125
- Dark and Spicy Pumpkin Loaf, 178, *179*, 180
- Molasses Spice Cookies, 125
- Panforte Nero, 184
- Pecan Spice Cookies, 127
- Spiced Apricots, 302
- Spicy Basil Sablés, 129
- Spicy Chocolate Sablés, 251
- Sweet Pickled Peaches, 306

Strawberry-Blueberry Tart, 275–76
Strawberry-Mint Salad, 60, *61*, 62–63
Strawberry Tartlets, *272*, 273–74
sugar, 39–40
- Cinnamon Sugar, 290
- Nutmeg Sugar, 283

Sweet Potato Cake, 101

Tangy Aromatic Crackers, *195*, 259
tarts:
- Chestnut Jam Tart, 212, *213*
- Chocolate Coconut Tart, 318
- Coconut and Mango Tart with Coconut Pastry Cream, 315
- Coconut Key Lime Tart, *316*, 317
- Lemon Tart, 154–55
- Mocha Cream Tart with Chocolate Crust, 247–48, *249*
- Sorghum Pecan Tart, 277, *278*, 279
- sticking to pan, 211
- Strawberry-Blueberry Tart, 275–76
- Strawberry Tartlets, *272*, 273–74
- Walnut and Honey Tart with Chestnut Crust, *208*, 209–11
- Yogurt Tart, 110, *111*, 112

Thai rice flour, 27, 28, 52, 53, 352, 354
tips for success, 48
Toffee Sauce, *182*, 183
tortes:
- Black Cherry Chocolate Linzer Torte, 245–46
- Carrot Coconut Almond Torte, 307–8
- Chocolate Chestnut Soufflé Cake, 206–7, *207*
- Coconut Chocolate Pecan Torte, 310, *311*

Triple Chocolate Layer Cake, 234–36
tuiles:
- Almond Tuiles, 67, *68*, 69
- Buckwheat Walnut or Hazelnut Tuiles, 188
- Oat and Almond Tuiles, 119
- Oat and Coconut Tuiles, 119
- shaping, 69

vanilla, 40
- The New Vanilla Pastry Cream, 342
- Silky Vanilla Pudding, 78

waffles, 21-22; Basic Waffles, 23
walnuts, *see* nuts
Whipped Cream, 343, 345
Whipped Ganache Filling, 236
white chocolate, 33, 37

xanthan gum, 40

yogurt:
- Almond Butter Cake with Spiced Apricots, 300, *301*, 302
- Buckwheat Cake with Rose Apples, 172, *173*, 174
- Caramel Apple Upside-Down Cake, 96, *97*, 98
- Chocolate Layer Cake, 94–95
- Cinnamon Crumb Cake, 99
- Corn Flour and Cranberry Scones, *150*, 151
- Corn Flour Biscuits, 147
- Corn Flour Tea Cake with Currants and Pistachios, 148, *149*
- Fig and Anise Scones, 270, *271*
- Hazelnut Layer Cake with Dark Chocolate Frosting and Blackberry Preserves, 304
- Peach-Almond Cobbler, 319
- Pistachio Butter Cake with Sweet Pickled Peaches, 305–6
- Poppy Seed Pound Cake, *92*, 93
- Simple Scones, 107
- Souffléed Corn Flour and Yogurt Puddings with Cajeta, 164, *165*
- Teff Blini, 258
- Ultimate Butter Cake, 90–91
- Yogurt Tart, 110, *111*, 112

CONVERSION CHARTS

Here are rounded-off equivalents between the metric system and the traditional systems that are used in the United States to measure weight and volume.

FRACTIONS	DECIMALS
⅛	.125
¼	.25
⅓	.33
⅜	.375
½	.5
⅝	.625
⅔	.67
¾	.75
⅞	.875

WEIGHTS

US/UK	METRIC
¼ oz	7 g
½ oz	15 g
1 oz	30 g
2 oz	55 g
3 oz	85 g
4 oz	110 g
5 oz	140 g
6 oz	170 g
7 oz	200 g
8 oz (½ lb)	225 g
9 oz	250 g
10 oz	280 g
11 oz	310 g
12 oz	340 g
13 oz	370 g
14 oz	400 g
15 oz	425 g
16 oz (1 lb)	455 g

VOLUME

AMERICAN	IMPERIAL	METRIC
¼ tsp		1.25 ml
½ tsp		2.5 ml
1 tsp		5 ml
½ Tbsp (1½ tsp)		7.5 ml
1 Tbsp (3 tsp)		15 ml
¼ cup (4 Tbsp)	2 fl oz	60 ml
⅓ cup (5 Tbsp)	2½ fl oz	75 ml
½ cup (8 Tbsp)	4 fl oz	125 ml
⅔ cup (10 Tbsp)	5½ fl oz	150 ml
¾ cup (12 Tbsp)	6 fl oz	175 ml
1 cup (16 Tbsp)	8 fl oz	250 ml
1¼ cups	10 fl oz	300 ml
1½ cups	12 fl oz	350 ml
2 cups (1 pint)	16 fl oz	500 ml
2½ cups	20 fl oz (1 pint)	625 ml
5 cups	40 fl oz (1 qt)	1.25 l

OVEN TEMPERATURES

	°F	°C	GAS MARK
very cool	250–275	130–140	½–1
cool	300	148	2
warm	325	163	3
moderate	350	177	4
moderately hot	375–400	190–204	5–6
hot	425	218	7
very hot	450–475	232–245	8–9

ALICE MEDRICH has won more cookbook-of-the-year awards and best in the dessert and baking category awards than any other author. She received her formal training at the prestigious École Lenôtre in France, and is credited with popularizing chocolate truffles in the United States when she began making and selling them at her influential Berkeley dessert shop, Cocolat. She has since left the retail world, devoting much of her career to teaching and sharing her expansive knowledge about baking (check out her online baking course on craftsy.com). Find her at AliceMedrich.com, and follow her on Twitter @Alice Medrich.